Homework Helpers

Eureka Math
Grade K

Special thanks go to the Gordan A. Cain Center and to the Department of Mathematics at Louisiana State University for their support in the development of *Eureka Math*.

Published by the non-profit Great Minds

Copyright © 2015 Great Minds. No part of this work may be reproduced, sold, or commercialized, in whole or in part, without written permission from Great Minds. Non-commercial use is licensed pursuant to a Creative Commons Attribution-NonCommercial-ShareAlike 4.0 license; for more information, go to http://greatminds.net/maps/math/copyright. "Great Minds" and "Eureka Math" are registered trademarks of Great Minds.

Printed in the U.S.A.
This book may be purchased from the publisher at eureka-math.org
10 9 8 7 6 5 4 3 2 1
ISBN 978-1-63255-823-7

Homework Helpers

Grade K
Module 1

GK-M1-Lesson 1

Color the things that are exactly the same. Color them so that they look like each other.

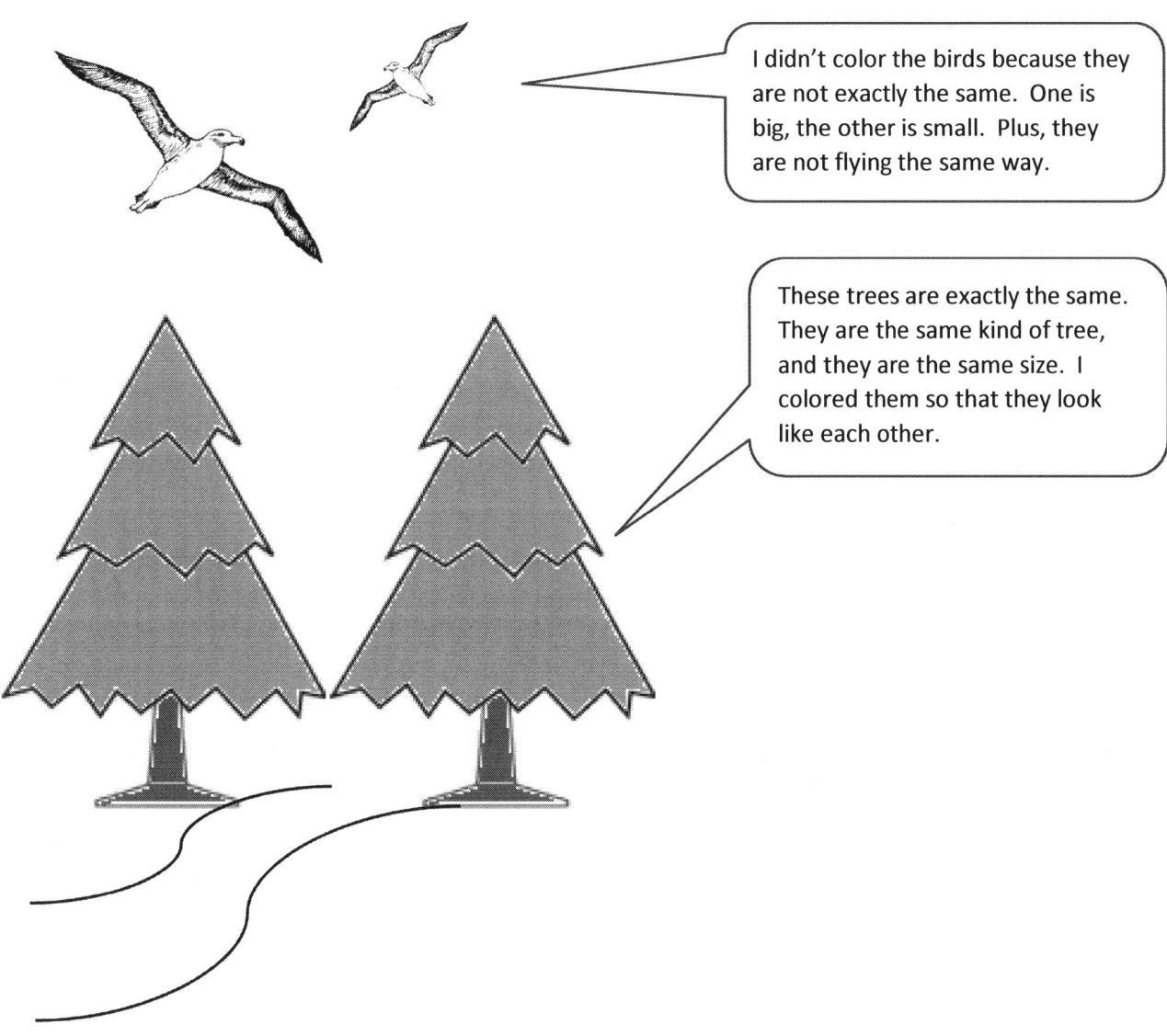

I didn't color the birds because they are not exactly the same. One is big, the other is small. Plus, they are not flying the same way.

These trees are exactly the same. They are the same kind of tree, and they are the same size. I colored them so that they look like each other.

Lesson 1: Analyze to find two objects that are *exactly the same* or *not exactly the same*.

GK-M1-Lesson 2

Draw a line between two objects that match. Use your words. "These are the same, but this one is _____, and this one is _____."

These are the same, but this one has spots on it, and this one doesn't.

GK-M1-Lesson 3

Make a picture of 2 things you use together. Tell why.

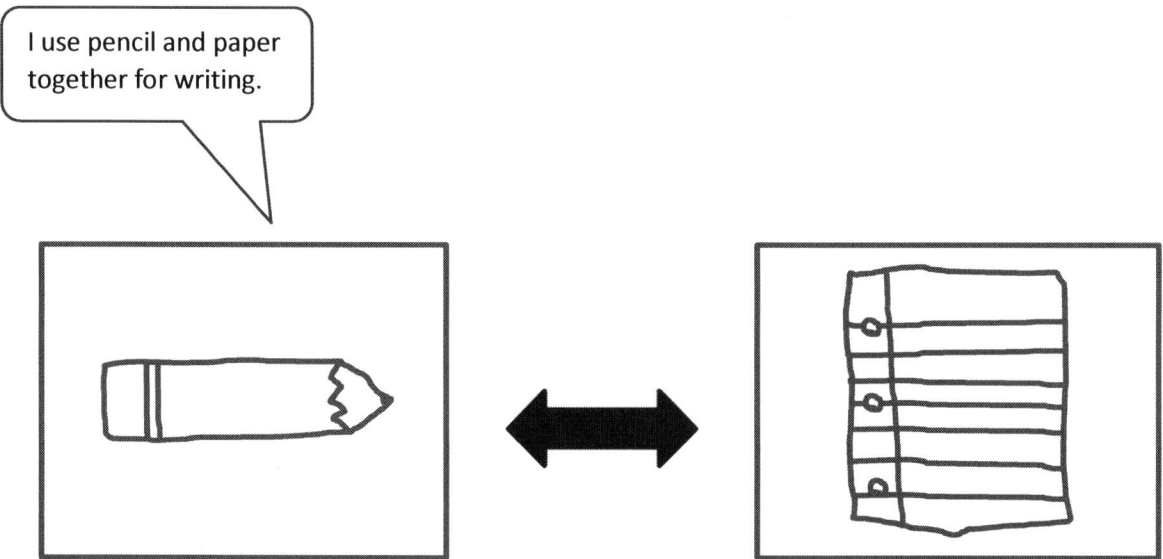

I use pencil and paper together for writing.

GK-M1-Lesson 4

Make two groups. Circle the things that belong to one group. Underline the things that belong to the other group. Tell someone why the items in each group belong together. (There is more than one way to make groups!)

I sorted them into two groups: stuffed animals and real animals. How did you sort them?

GK-M1-Lesson 5

Use the cutouts. Glue the pictures to show where each belongs. Tell an adult how many are in each place.

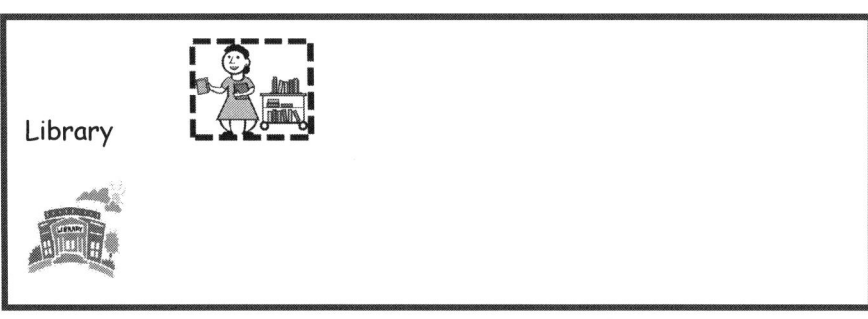

A lemon, a pineapple, and a shopping cart belong in the grocery store.

There are 3 grocery store things.

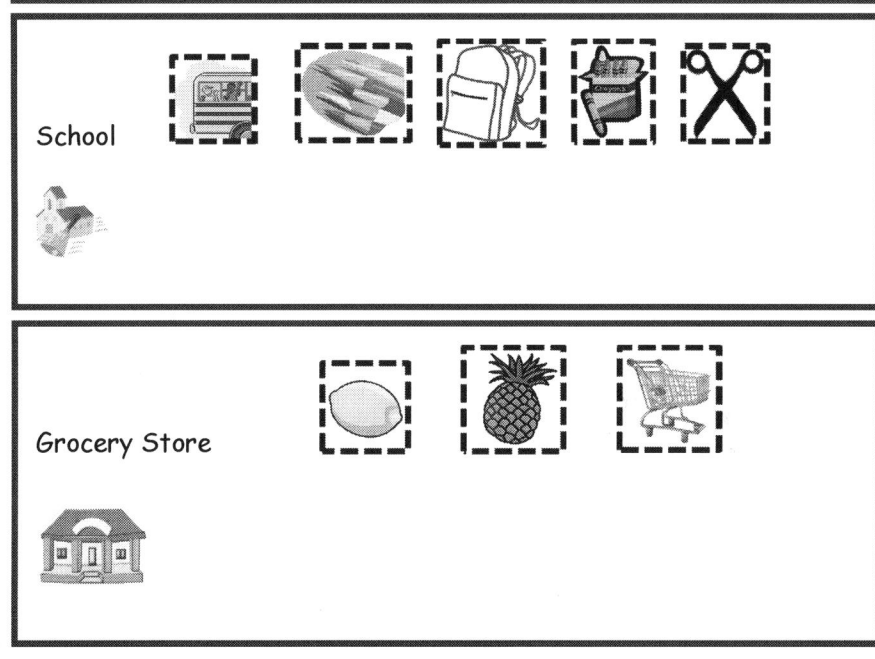

Lesson 5: Classify items into three categories, determine the count in each, and reason about how the last number named determines the total.

GK-M1-Lesson 6

Draw lines to put the treasures in the boxes.

I can sort by count!
Groups of 2 belong in the 2 box.
Groups of 3 belong in the 3 box.
Groups of 4 belong in the 4 box.

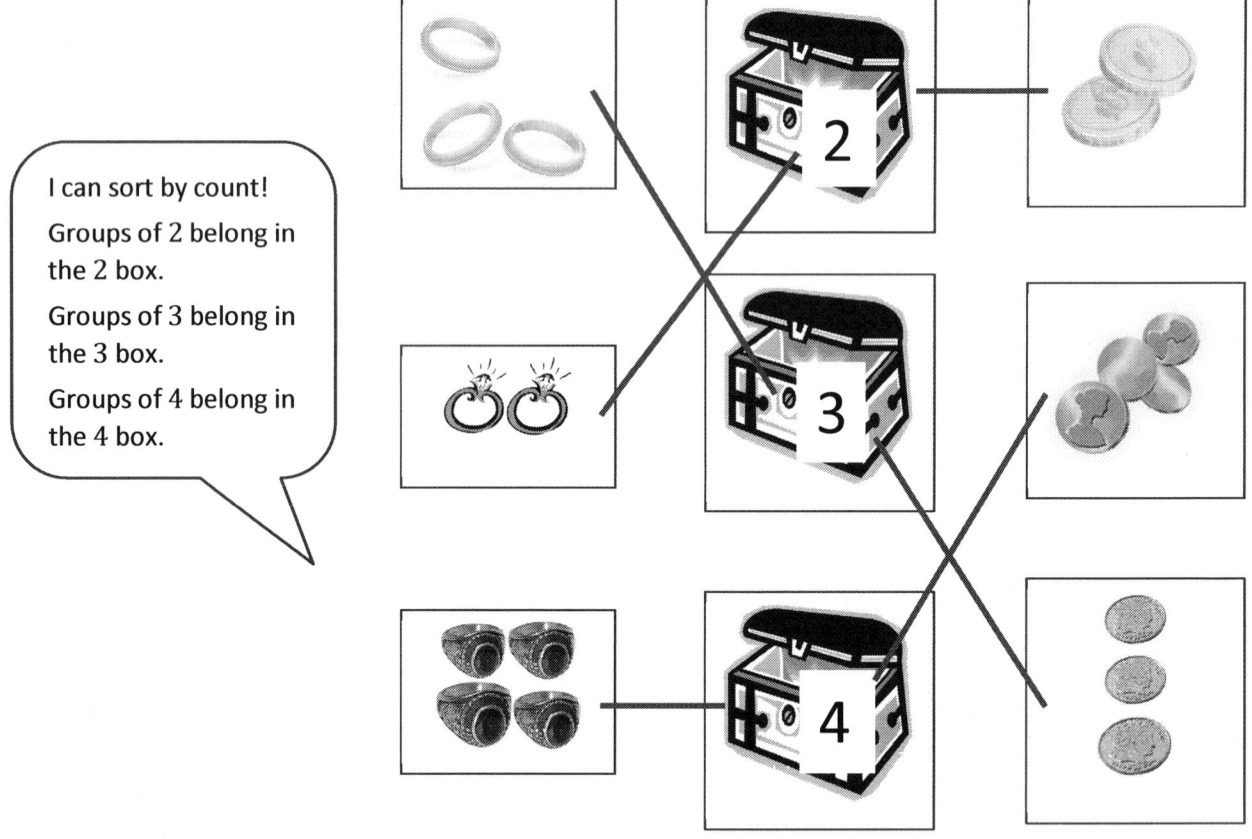

Lesson 6: Sort categories by count. Identify categories with 2, 3, and 4 within a given scenario.

GK-M1-Lesson 7

Count and color.

> I ask for help reading the words. Then I color in the boxes to make a color code.

> I see 2 of these. I will color them blue, just like the card.

GK-M1-Lesson 8

Count. Circle the number that tells *how many*.

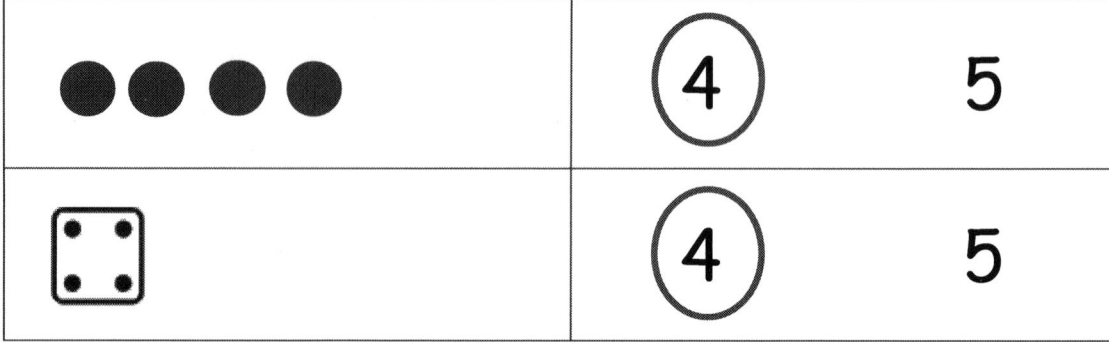

This one is easy! I counted 4 dots in a straight line. So I circle 4.

I counted 4 this time, too, but it looks different. I see 2 on the top and 2 on the bottom.

Lesson 8: Answer *how many* questions to 5 in linear configurations (5-group), with 4 in an array configuration. Compare ways to count five fingers.

Homework Helper

A Story of Units

GK-M1-Lesson 9

Count the circles, and box the correct number. Color in the same number of circles on the right as the shaded ones on the left to show hidden partners.

There are 4 circles: 3 of them are gray, and 1 is white. The hidden partners are 3 and 1.

3 |4| 5

I color in 3 circles.
I see 3 and 1 hiding inside of 4.

Lesson 9: Within linear and array dot configurations of numbers 3, 4, and 5, find hidden partners.

GK-M1-Lesson 10

Count how many. Draw a box around that number. Then, color 1 of the circles in each group.

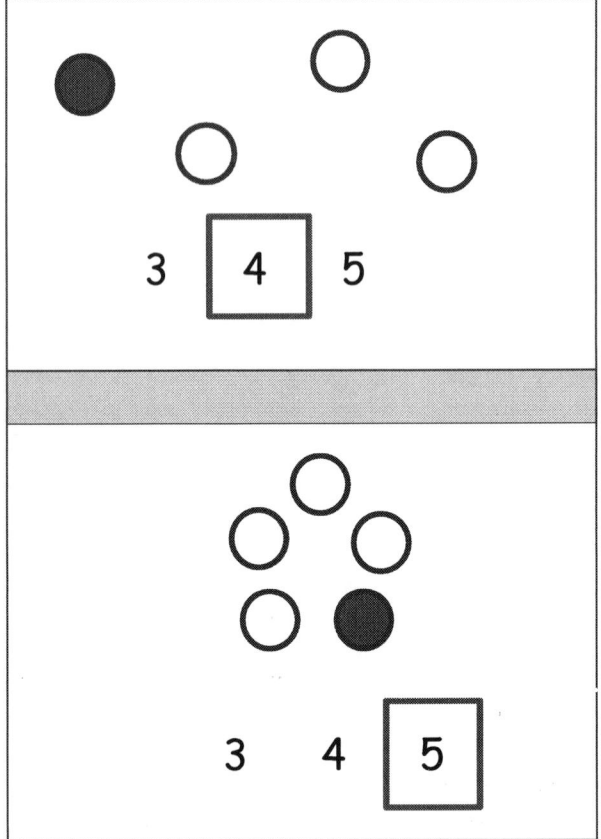

There are 4 circles.
I color 1 of them.
The hidden partners are 3 and 1.

I color in 1 circle.
I see 4 and 1 hiding inside of 5.

Homework Helper — A Story of Units K•1

GK-M1-Lesson 11

Color the shapes to show 1 + 2. Use your 2 favorite colors.

I color 1 blue and 2 red.
3 is the same as 1 and 2.

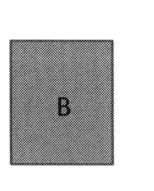
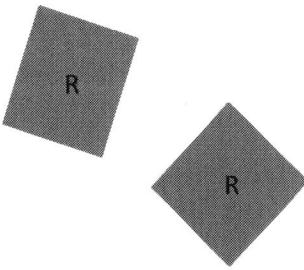

How many shapes are there?

Circle the number.   1   2   ③   4   5

Lesson 11: Model decompositions of 3 with materials, drawings, and expressions. Represent the decomposition as 1 + 2 and 2 + 1.

GK-M1-Lesson 12

How many? Draw a line between each picture and its number.

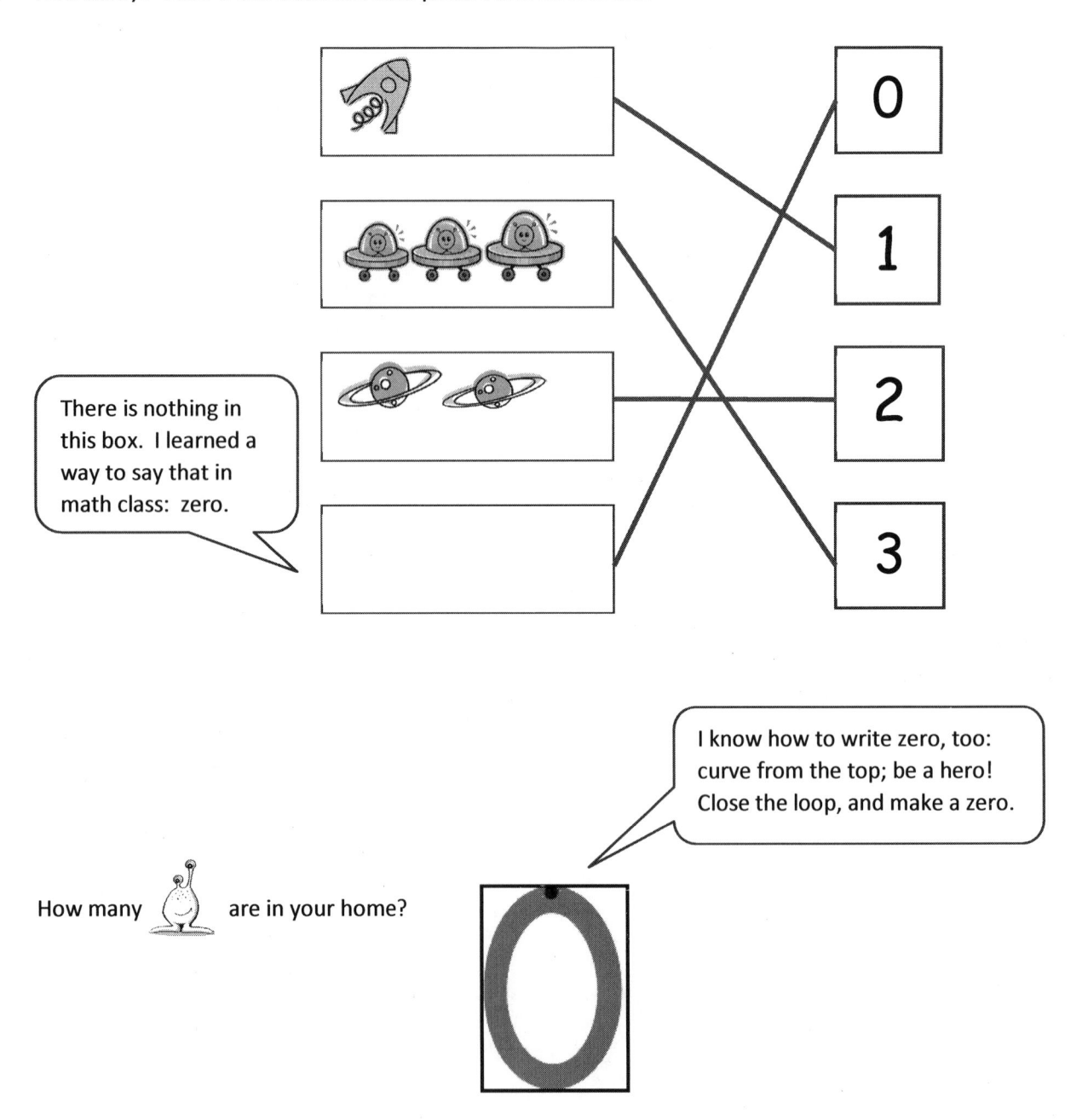

> There is nothing in this box. I learned a way to say that in math class: zero.

> I know how to write zero, too: curve from the top; be a hero! Close the loop, and make a zero.

How many 👽 are in your home?

Lesson 12: Understand the meaning of zero. Write the numeral 0.

Homework Helper — A Story of Units

GK-M1-Lesson 13

Count the objects. Write how many.

1, 2.
I count 2 cats.
I write the number 2.

Write the missing numbers.

1, 2, **3**

3, 2, 1, **0**

Lesson 13: Order and write numerals 0–3 to answer *how many* questions.

GK-M1-Lesson 14

Color the stars so that 1 is yellow and 2 are red.

> I count 3 things. I color 1 star yellow and 2 stars red. When I take apart 3, its parts are 2 and 1.

There are  stars.

> I read the number sentence like this: 3 is the same as 1 and 2.

$$\boxed{3} = 1 + 2$$

Lesson 14: Write numerals 1–3. Represent decompositions with materials, drawings, and equations, $3 = 2 + 1$ and $3 = 1 + 2$.

Homework Helper — A Story of Units

GK-M1-Lesson 15

Count the shapes and write the numbers. Mark each shape as you count.

These fruits are everywhere! I mark each one as I count. That way, I don't count the same one twice.

1, 2, 3, 4. There are 4 watermelons.

1, 2, 3, 4, 5. There are 5 oranges.

I can write 4. Trace down the side; cross the middle for fun. Top to bottom, and you are done!

I can write 5. Trace down the side; curve like that. Back to the dot, and give it a hat!

How many?

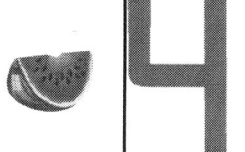

Lesson 15: Order and write numerals 4 and 5 to answer *how many* questions in categories; sort by count.

Homework Helper — A Story of Units

GK-M1-Lesson 16

Write the missing numbers:

1, 2, **3**, 4, **5**

> I can count up and down. Counting out loud helps me find the missing number.

4, 3, 2, 1, **0**

Draw 3 yellow fish and 2 green fish.

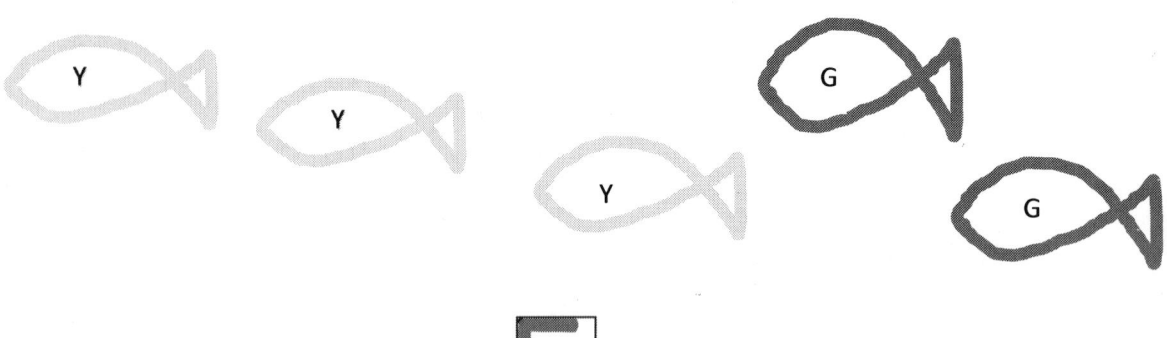

How many fish are there in all? There are **5** fish.

3 fish and 2 fish make **5** fish.   5 is the same as **3** and **2**.

> Breaking apart 5 is easy. I see 3 and 2 in my picture.

> I can put together 3 and 2 to make 5.

GK-M1-Lesson 17

Color 4.

Color 6.

Connect the boxes with the same number.

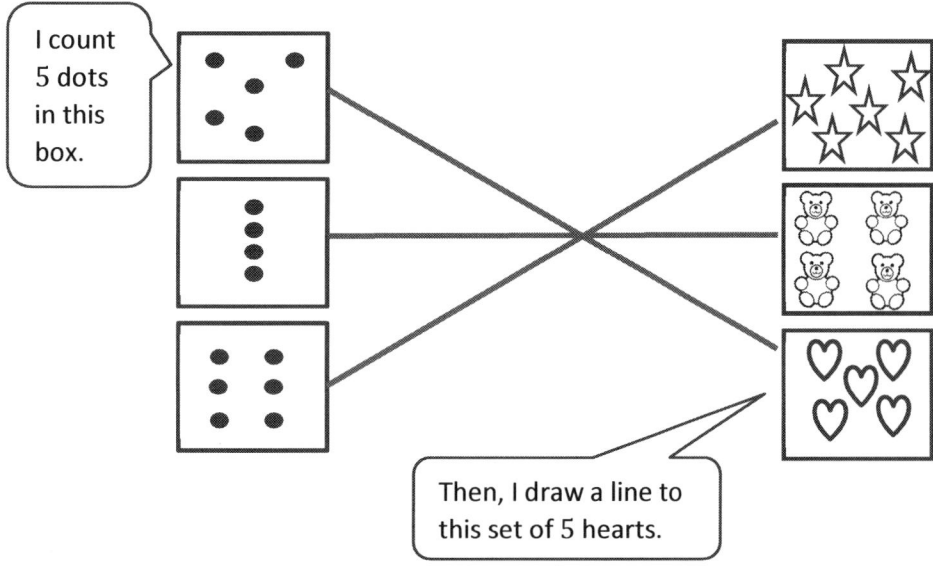

GK-M1-Lesson 18

Color 4.

I can count stars in a circle! I color 4 stars. There are 2 stars left. That makes 6 stars in all.

It's easy for me to count objects in a row. I count 7 balloons!

Circle 5  balloons.

When I circle 5 balloons, I notice 2 balloons are left.

   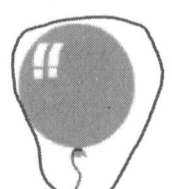   

Homework Helper — A Story of Units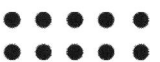

GK-M1-Lesson 19

• • • • •

5-group

Like fingers on a hand, we can make groups of 5 (and some more).

• • • • •   • • • • •   • • • • •   • • • • •   • • • • •
•   • •   • • •   • • • •   • • • • •

Draw a line from the numeral to the 5-group it matches.

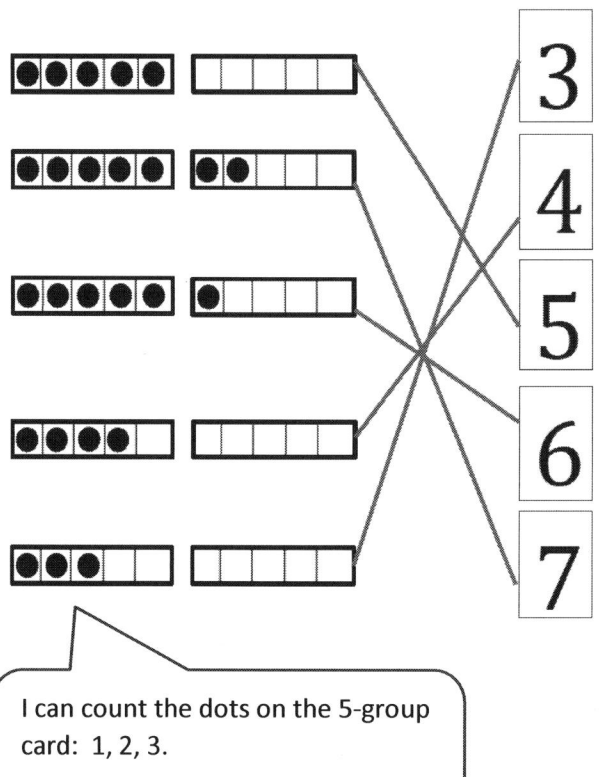

I can count the dots on the 5-group card: 1, 2, 3.
With practice, I'll be able to just see that there are three.

Here's one card with 5 and another with 2. I can count 5, 6, 7.
Or, I can count them all. That's seven!

Lesson 19: Count 5–7 linking cubes in linear configurations. Match with numeral 7. Count on fingers from 1 to 7, and connect to 5-group images.

Fill in the missing numbers.

> I count up to 7, starting from any number. Look at me! I can write my numbers!

4, 5, 6, 7

7, 6, 5, 4, 3, 2

1, 2, 3, 4, 5, 6, 7

Homework Helper — A Story of Units

GK-M1-Lesson 20

How many? Write the number in the box.

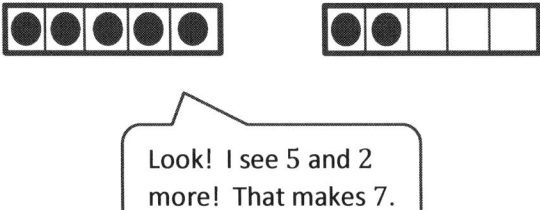

Look! I see 5 and 2 more! That makes 7.

Count how many. Write the number in the box.

Draw a line to show how you counted the triangles.

I can count the triangles! Here is my counting path. What's yours?

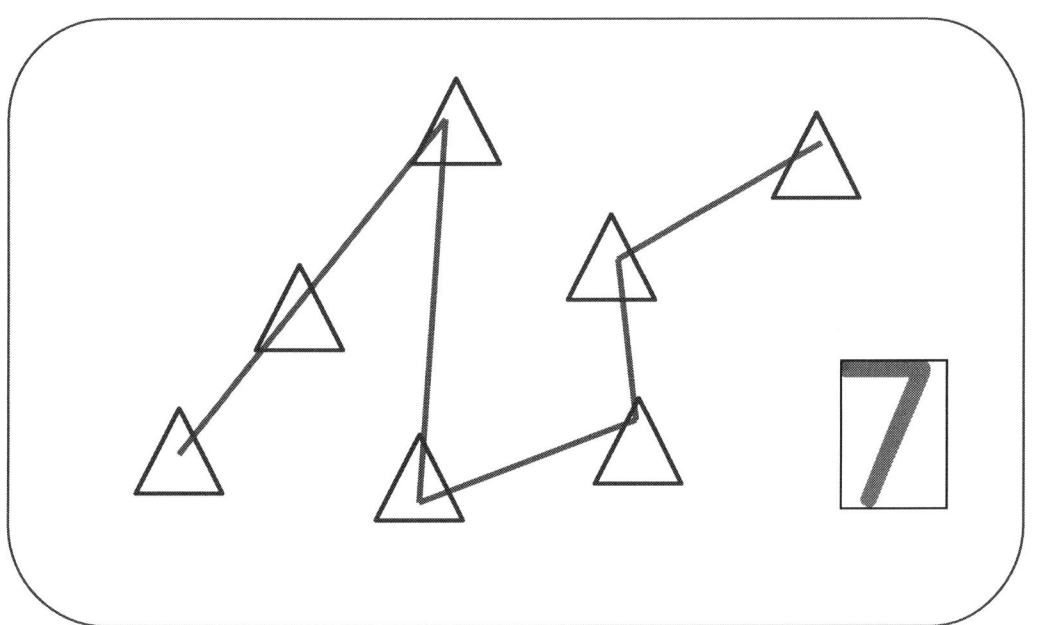

There are 7 in all! "A straight line and down from heaven; that's the way we make a 7."

Lesson 20: Reason about sets of 7 varied objects in circular and scattered configurations. Find a path through the scattered configuration. Write numeral 7. Ask, "How is your seven different from mine?"

Homework Helper — A Story of Units — K•1

GK-M1-Lesson 21

Color 4 ladybugs red. Color 4 ladybugs yellow.
Count and circle how many.

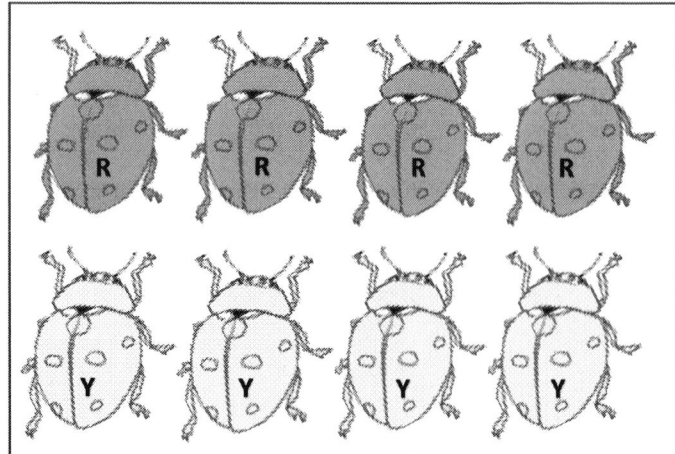

> These two rows have the same number of ladybugs. I can see 4 and 4 hiding in 8.

6   7   ⑧

Color 4 ladybugs blue. Color 4 ladybugs orange.
Count and circle how many.

> It doesn't matter whether the ladybugs are arranged in rows or columns; there are still 8 ladybugs in all!

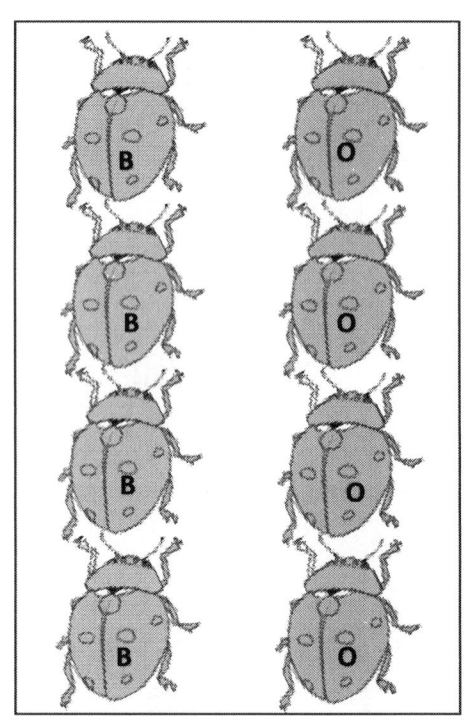

Lesson 21: Compare counts of 8. Match with numeral 8.

Homework Helper

A Story of Units — K•1

Count how many. Write the number in the box.

> I see 5 and 2 hiding in 7.

7

Lesson 21: Compare counts of 8. Match with numeral 8.

GK-M1-Lesson 22

Draw 8 beads around the circle.

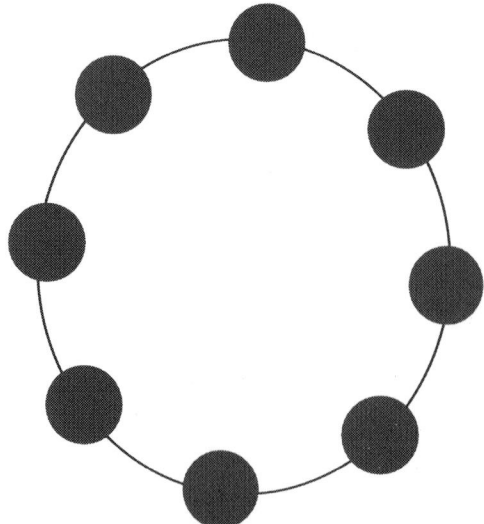

How did you count? What is your strategy?

Color 8. Draw a line to show your counting path.

This path shows how I counted the stars.
How did you count?

Lesson 22: Arrange and strategize to count 8 beans in circular (around a cup) and scattered configurations. Write numeral 8. Find a path through the scattered set, and compare paths with a partner.

Homework Helper — A Story of Units

Count how many. Write the number in the box

I can write 8. Make an S, and do not stop. Go right back up, and an 8 you've got!

Lesson 22: Arrange and strategize to count 8 beans in circular (around a cup) and scattered configurations. Write numeral 8. Find a path through the scattered set, and compare paths with a partner.

GK-M1-Lesson 23

Color 9 circles.

"I can see 5 and 4 hiding in 9."

"I can see 1 and 9 is ten!"

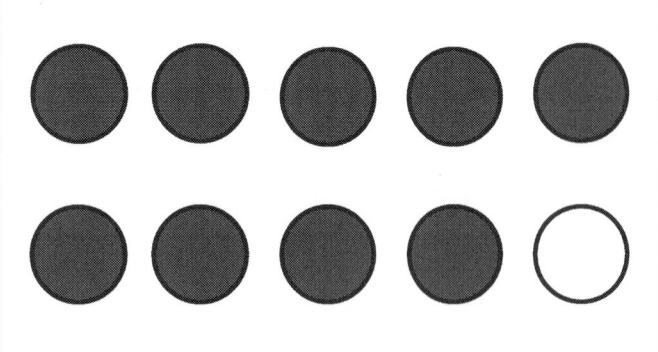

Draw 9 shapes.

"Do your shapes look like mine? There are so many ways to draw and arrange nine!"

Homework Helper — A Story of Units — K•1

GK-M1-Lesson 24

Color 9 circles.

Look at me! I can count 9 circles scattered about. I don't count any circles more than once. I have a strategy. Do you?

Count. Write the number in the box.

I remember how to write 9. A hoop and a line. That's the way we make nine!

Lesson 24: Strategize to count 9 objects in circular (around a paper plate) and scattered configurations printed on paper. Write numeral 9. Represent a path through the scatter count with a pencil. Number each object.

GK-M1-Lesson 25

Color 5 suns. Color 5 more suns a different color.

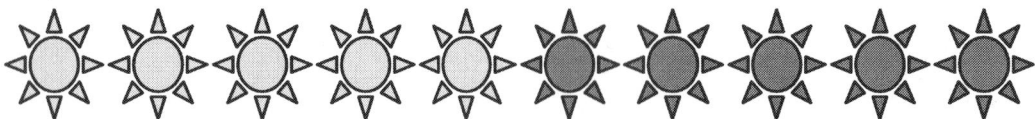

I count 10 in all! Ten is the same as 5 and 5.

Color 9 stars. Color 1 more star a different color.

I count 10 stars in all! Nine and 1 more make ten!

I see 2 columns of 5. I see 5 rows of 2. They both show 10 in all.

Draw 5 circles under the row of circles. Color 5 circles yellow. Color 5 circles green.

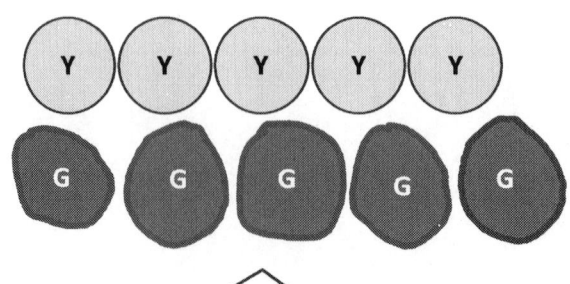

I color 1 row of 5 circles. I can draw 5 more circles. Look at my 2 rows of five!

Homework Helper — A Story of Units

GK-M1-Lesson 26

Draw 5 circles in a row. Draw another 5 circles in a row under them.
How many circles did you draw?

Write the number in the box.

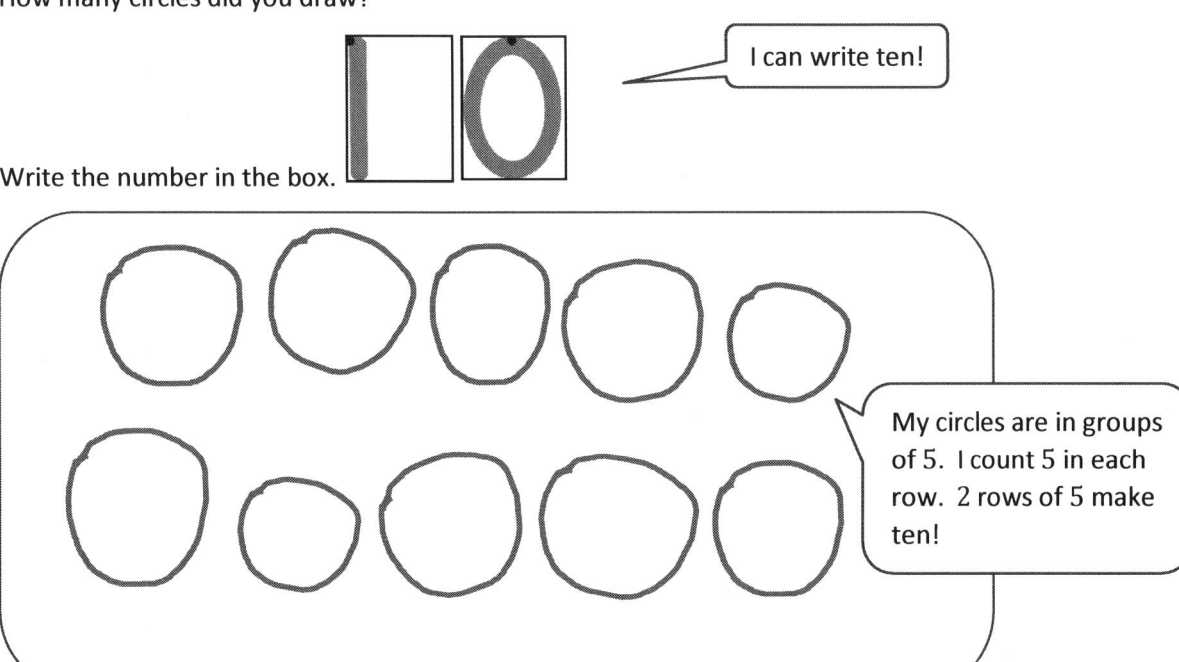

"I can write ten!"

"My circles are in groups of 5. I count 5 in each row. 2 rows of 5 make ten!"

Write how many in the box.

"These triangles are not arranged in a line. But, I can count them all without counting twice. I've got a strategy!"

Lesson 26: Count 10 objects in linear and array configurations (2 fives). Match with numeral 10. Place on the 5-group mat. Dialogue about 9 and 10. Write numeral 10.

Homework Helper — A Story of Units

GK-M1-Lesson 27

Draw enough  to make 10.

I count 5 hearts. I can draw more to make 10.

I made 2 groups of five! Like my fingers on my 2 hands, altogether there are ten!

Draw enough  to make 10.

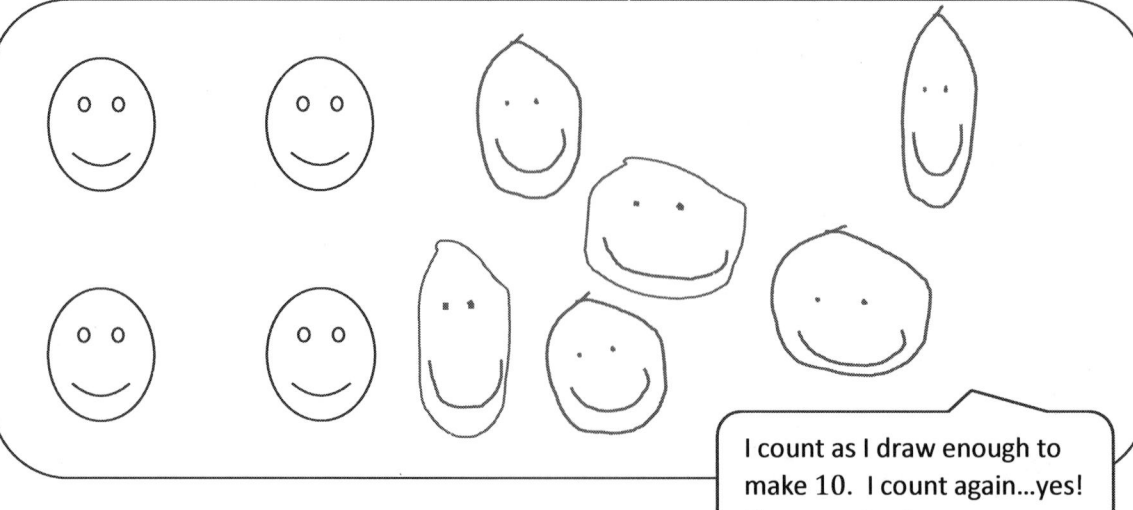

I count as I draw enough to make 10. I count again...yes! There are ten!

Homework Helper — A Story of Units

GK-M1-Lesson 28

Make up a story about 10 things in your house. Draw a picture to go with your story. Be ready to share your story at school tomorrow.

> I remember math stories we acted out in class today. Stories like, "8 students. 4 are girls. How many are boys?"
> I can draw and tell a story. Can you solve?

> Mama and I ate a snack. There were 10 things on the table. Then, I dropped my fork on the floor. How many things are still on the table?

Lesson 28: Act out *result unknown* story problems without equations.

Homework Helper — A Story of Units K•1

GK-M1-Lesson 29

Count the dots. Write how many. Draw the same number of dots below but going up and down instead of across.

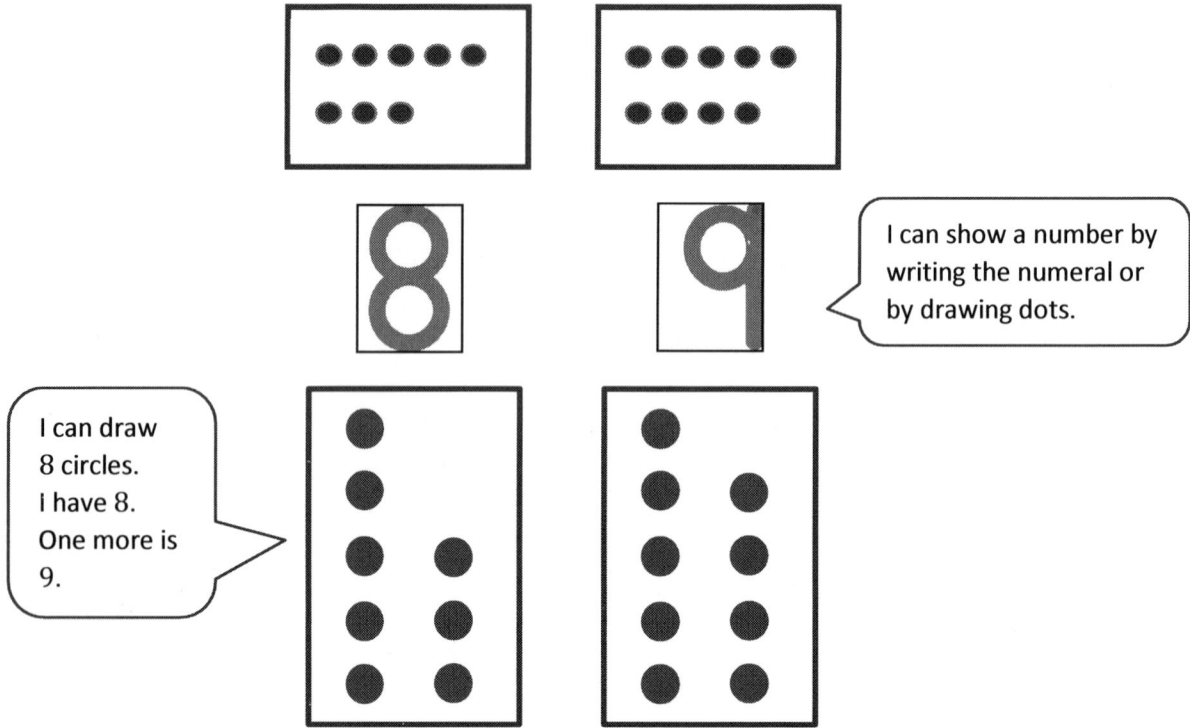

Make your own 5-group cards! Cut the cards out on the dotted lines. On one side, write the numbers from 1 to 10. On the other side, show the 5-group dot picture that goes with the number.

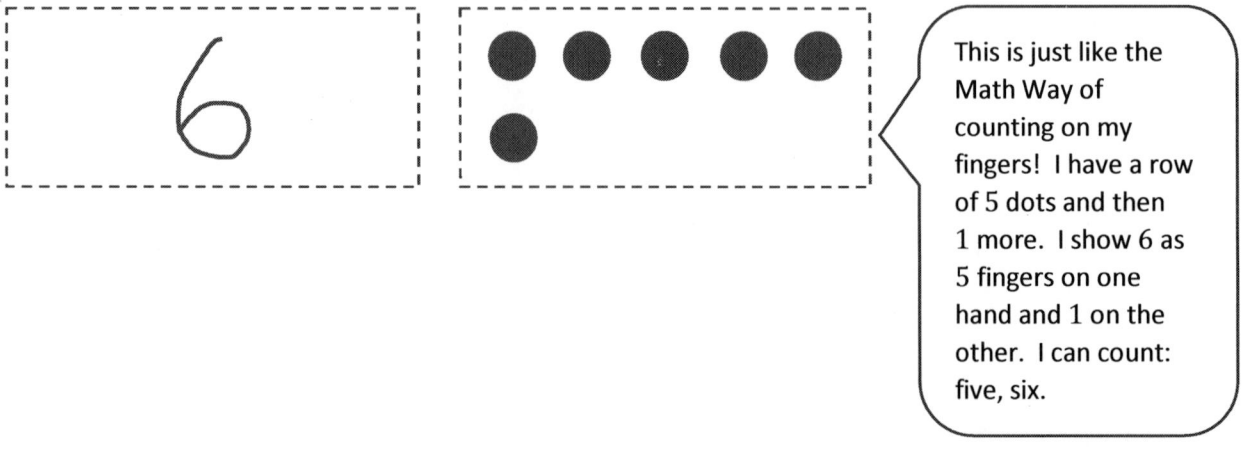

Lesson 29: Order and match numeral and dot cards from 1 to 10. State 1 more than a given number.

Homework Helper — A Story of Units

GK-M1-Lesson 30

Draw the missing stairs. Write the numbers below each step.

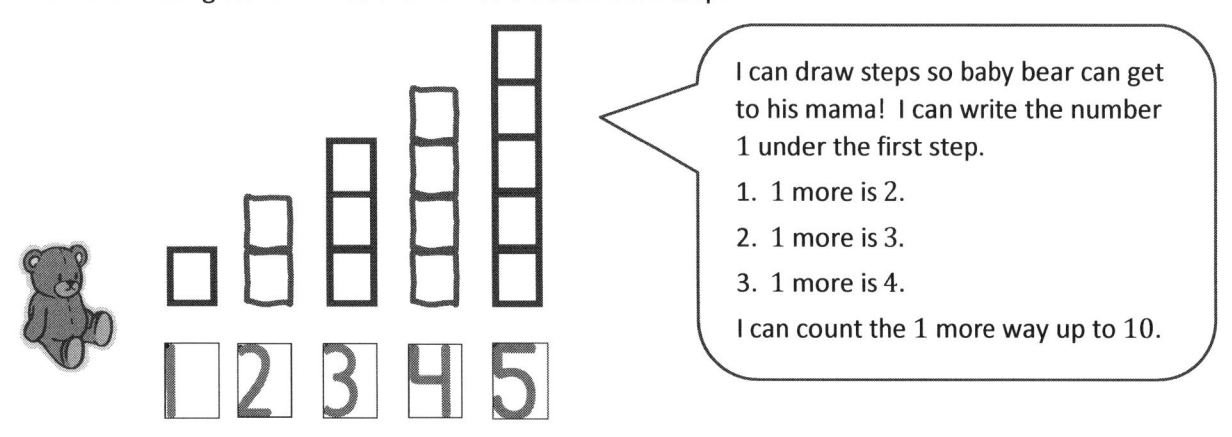

> I can draw steps so baby bear can get to his mama! I can write the number 1 under the first step.
> 1. 1 more is 2.
> 2. 1 more is 3.
> 3. 1 more is 4.
> I can count the 1 more way up to 10.

Draw 1 more cube on each stair so the cubes match the number. Say as you draw, "1. One more is two. 2. One more is three."

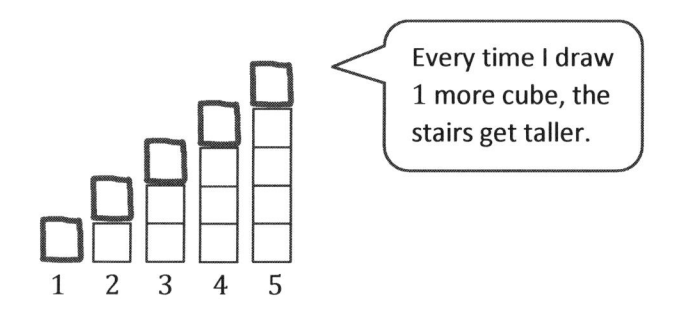

> Every time I draw 1 more cube, the stairs get taller.

Lesson 30: Make math stairs from 1 to 10 in cooperative groups.

GK-M1-Lesson 31

Draw one more circle. Color all the circles, and write how many.

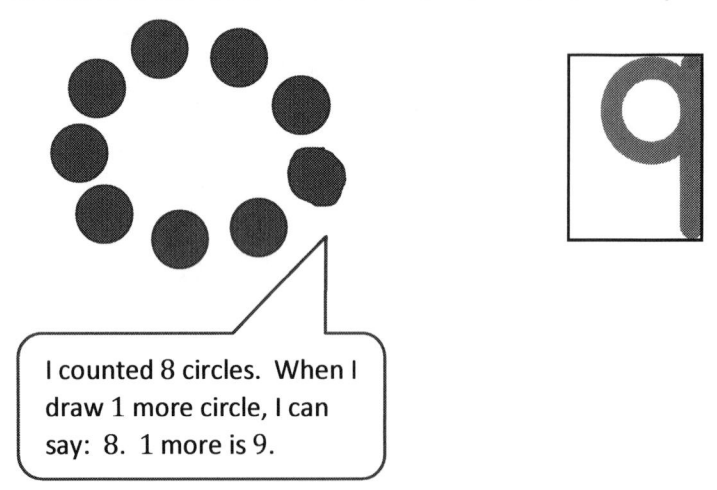

> I counted 8 circles. When I draw 1 more circle, I can say: 8. 1 more is 9.

Draw one more star. Color all the stars, and write how many.

> I counted 6 stars. Then, I can say: 6. 1 more is 7.

Lesson 31: Arrange, analyze, and draw 1 more up to 10 in configurations other than towers.

Homework Helper — A Story of Units

GK-M1-Lesson 32

Write the missing numbers.

2, **3**, **4**, 5, 6, **7**, **8**, **9**, 10

> Each number in the row is 1 more. 6. 1 more is 7. Then 8. Then 9.

Draw X's or O's to show 1 more.

```
XXXX        OOO
XX          O
```

> I don't have to start counting at 1 every time. I know there are 3 O's. 1 more is 4. If I drew the O's in a line, there would still be 4 of them.

Tell someone a story about "1 more…and then 1 more." Draw a picture about your story.

🍎🍎🍎   🍎   🍎

> Listen to my story: I have 3 apples in a basket. I put 1 more apple in my basket. 3. 1 more is 4. Then, I put 1 more in my basket. 4. 1 more is 5. I have 5 apples now!

Lesson 32: Arrange, analyze, and draw sequences of quantities of 1 more, beginning with numbers other than 1.

GK-M1-Lesson 33

Make 5-Group Cards: Cut the cards out on the dotted lines. On one side, write the numbers from 1-10. On the other side, show the 5-group dot picture that goes with the number. Mix up your cards, and practice putting them in order the "1 less way."

> When I put my cards in order from 10 to 1, I see a pattern. Each dot picture is 1 less, and each number is 1 less.

Homework Helper

A Story of Units

GK-M1-Lesson 34

Count and color the triangles. Draw a group of triangles that is 1 less. Write how many you drew.

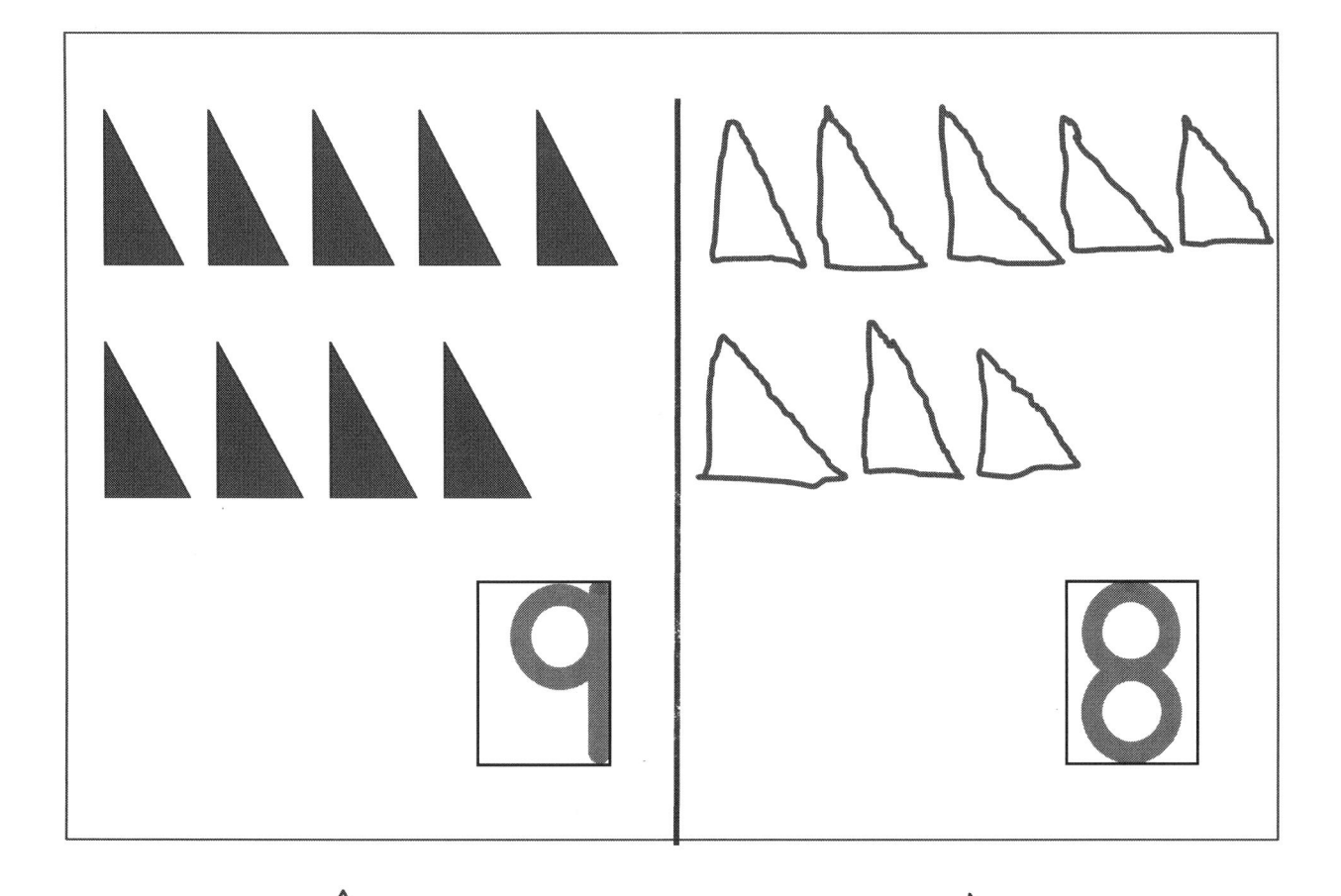

9 | 8

> I remember the 1 more pattern when we counted from 1 to 10. This is just the opposite! Now, I can count down from 10 to 1, and each number is 1 less!

> Look, one triangle has disappeared! 9. 1 less is 8. If I make another triangle disappear, I can say, 8. 1 less is 7.

Lesson 34: Count down from 10 to 1, and state 1 less than a given number.

Homework Helper — A Story of Units

GK-M1-Lesson 35

Count all the squares in each tower, and write how many. Share with someone what you notice!

> I can count the squares in this tower. There are 4. 1 less is 3.
> The towers keep getting smaller and so do the numbers!

> I can count the squares in the last tower, too. There is 1 square left. 1 less is 0.

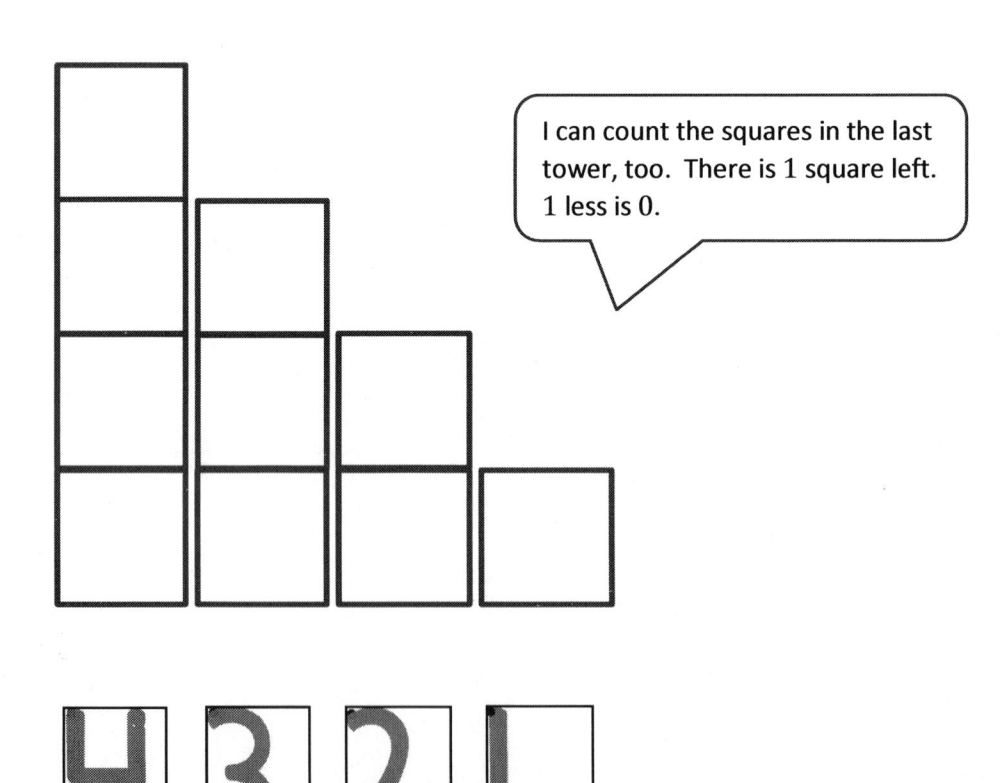

4 3 2 1

Lesson 35: Arrange number towers in order from 10 to 1, and describe the pattern.

Homework Helper   A Story of Units K•1

GK-M1-Lesson 36

Draw bracelets with the number of beads shown. Write the missing number. Hint: The missing number is 1 less!

> I had 8 beads. I know that 1 less is 7. I can call this my 7 bracelet! The next one will be my 6 bracelet. Each bracelet has 1 less.

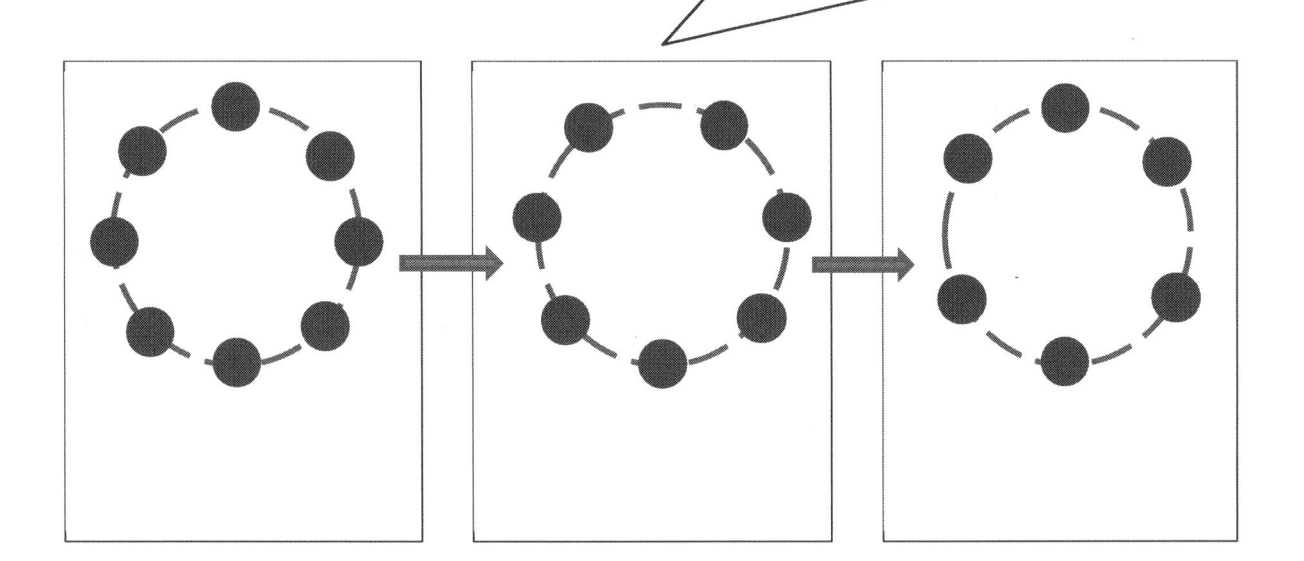

8   **7**   **6**

> I can count down from 10 to 0. When I start at 10, I know that the next number will be 1 less.

10, 9, **8**, **7**, 6, 5, 4, **3**, **2**, 1, 0

Lesson 36: Arrange, analyze, and draw sequences of quantities that are 1 less in configurations other than towers.

Homework Helper — A Story of Units K•1

GK-M1-Lesson 37

Note: Be sure to ask your child about his/her mystery number from today's Number Fair!

Count how many are in each group. Write the number in the box. Circle the smaller group.

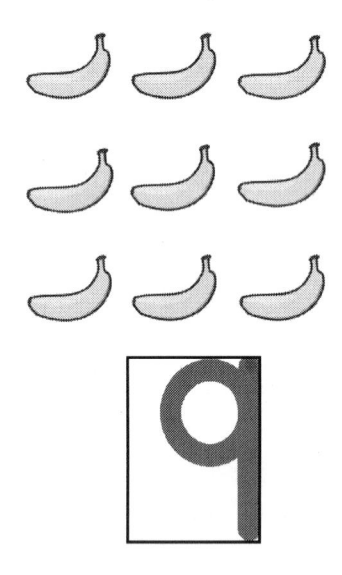

9 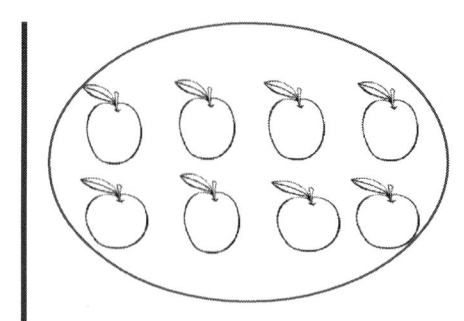 | **8**

> I see rows of bananas and apples. I can count 8 apples. I know that 8 is smaller than 9.
>
> I can say 9. 1 less is 8.
> Or I can say 8. 1 more is 9.

Draw some flowers.

How many? **7**

> I can draw 7 flowers in 5 groups. I can count them: fiiiive, six, seven. I know how to write the number 7.

Lesson 37: Culminating task

Homework Helpers

Grade K
Module 2

Homework Helper — A Story of Units — K•2

GK-M2-Lesson 1

Draw a line from the shape to its matching object.

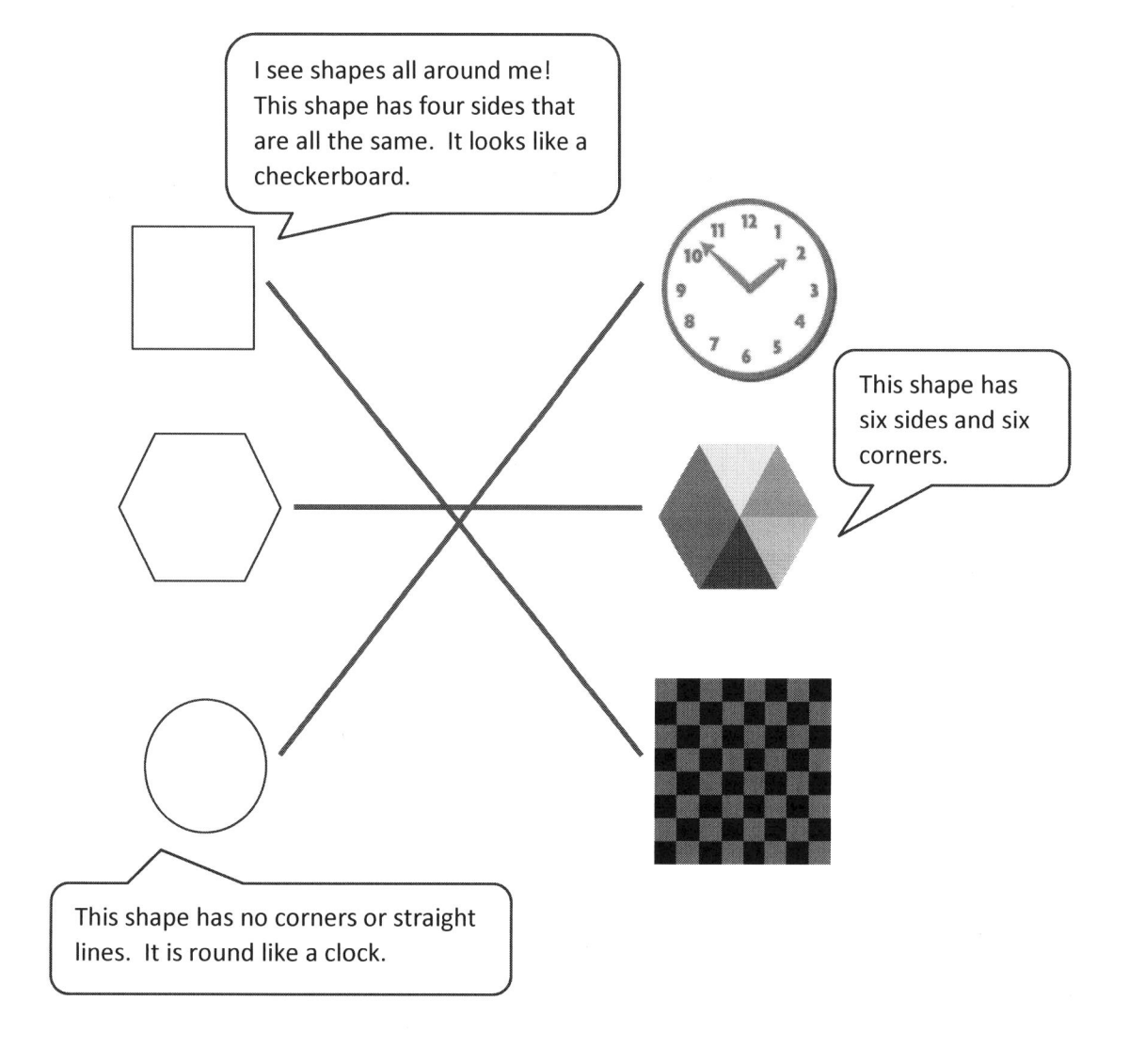

I see shapes all around me! This shape has four sides that are all the same. It looks like a checkerboard.

This shape has six sides and six corners.

This shape has no corners or straight lines. It is round like a clock.

Lesson 1: Find and describe flat triangles, squares, rectangles, hexagons, and circles using informal language without naming.

GK-M2-Lesson 2

Color the triangles red and the other shapes blue.

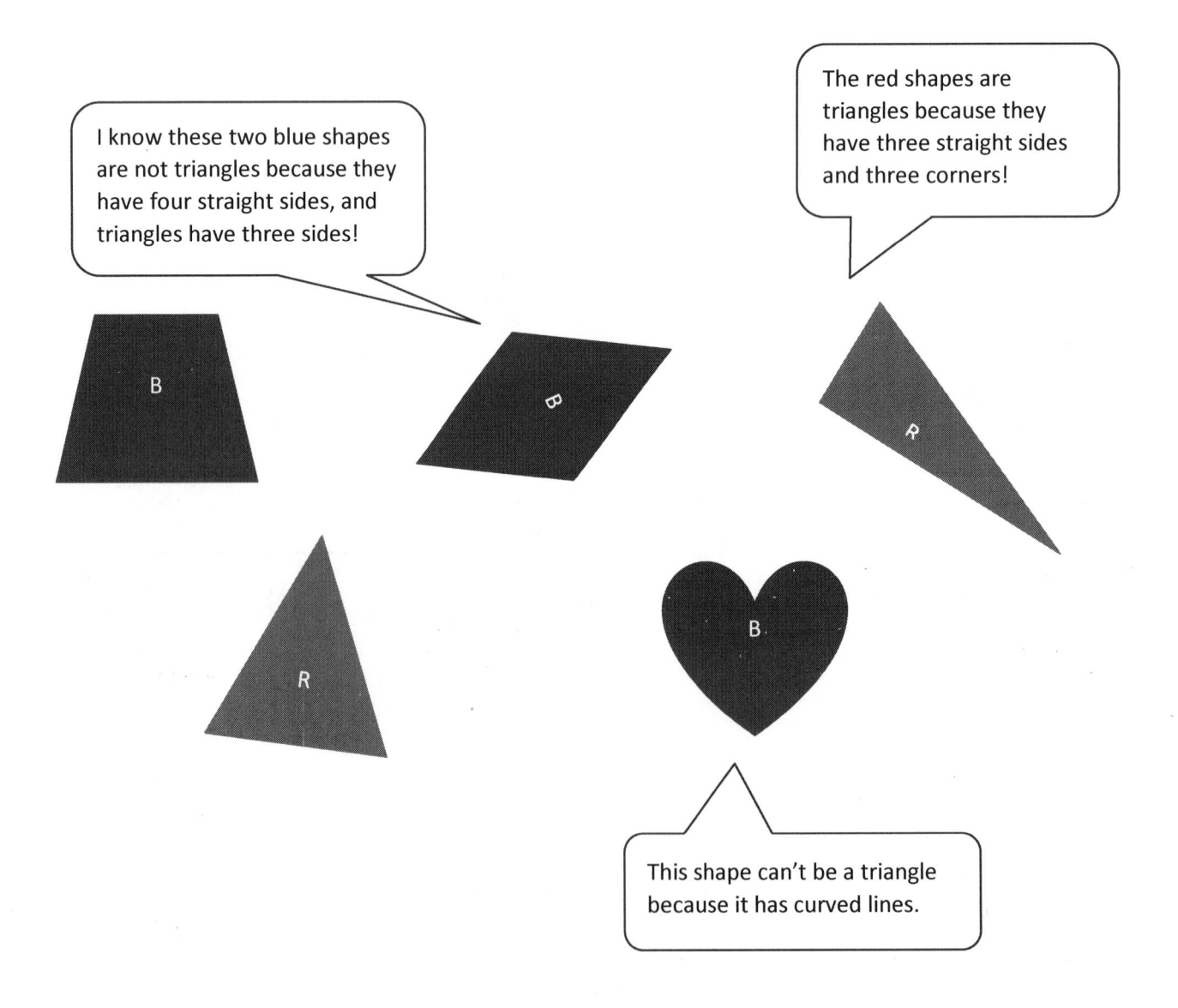

I know these two blue shapes are not triangles because they have four straight sides, and triangles have three sides!

The red shapes are triangles because they have three straight sides and three corners!

This shape can't be a triangle because it has curved lines.

GK-M2-Lesson 3

Color all the rectangles red. Color all the triangles green.

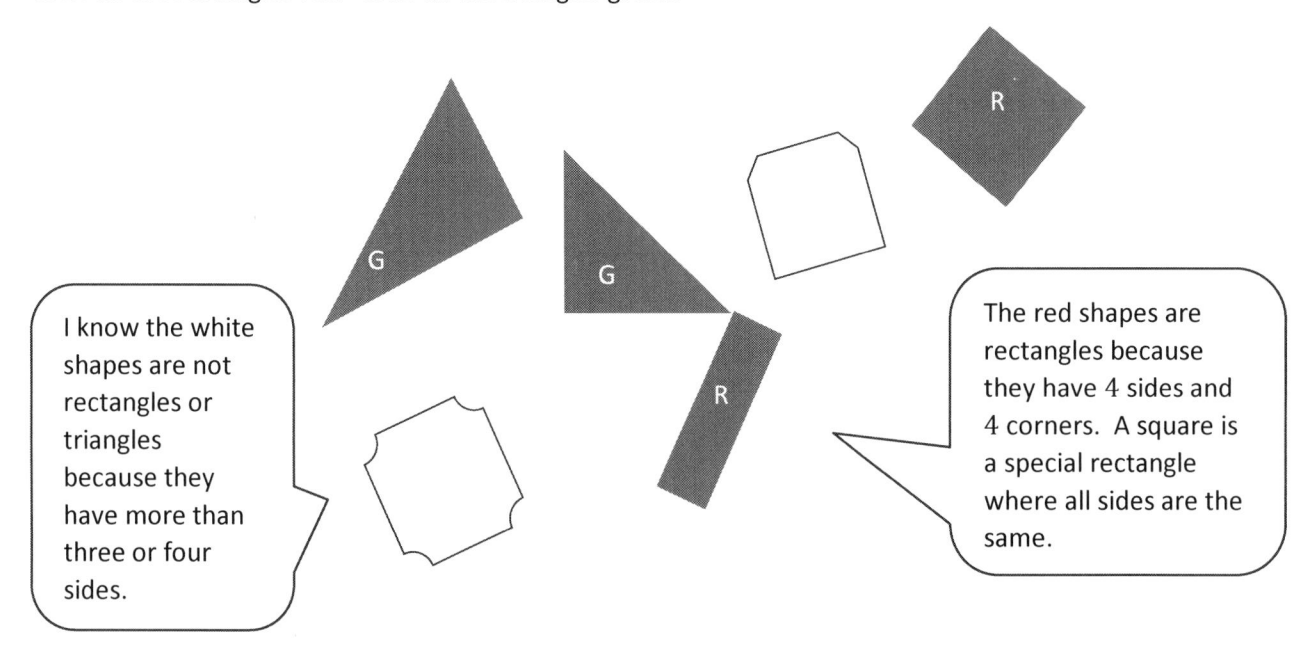

I know the white shapes are not rectangles or triangles because they have more than three or four sides.

The red shapes are rectangles because they have 4 sides and 4 corners. A square is a special rectangle where all sides are the same.

In the box, draw 2 rectangles and 2 triangles. How many shapes did you draw? Put your answer in the circle.

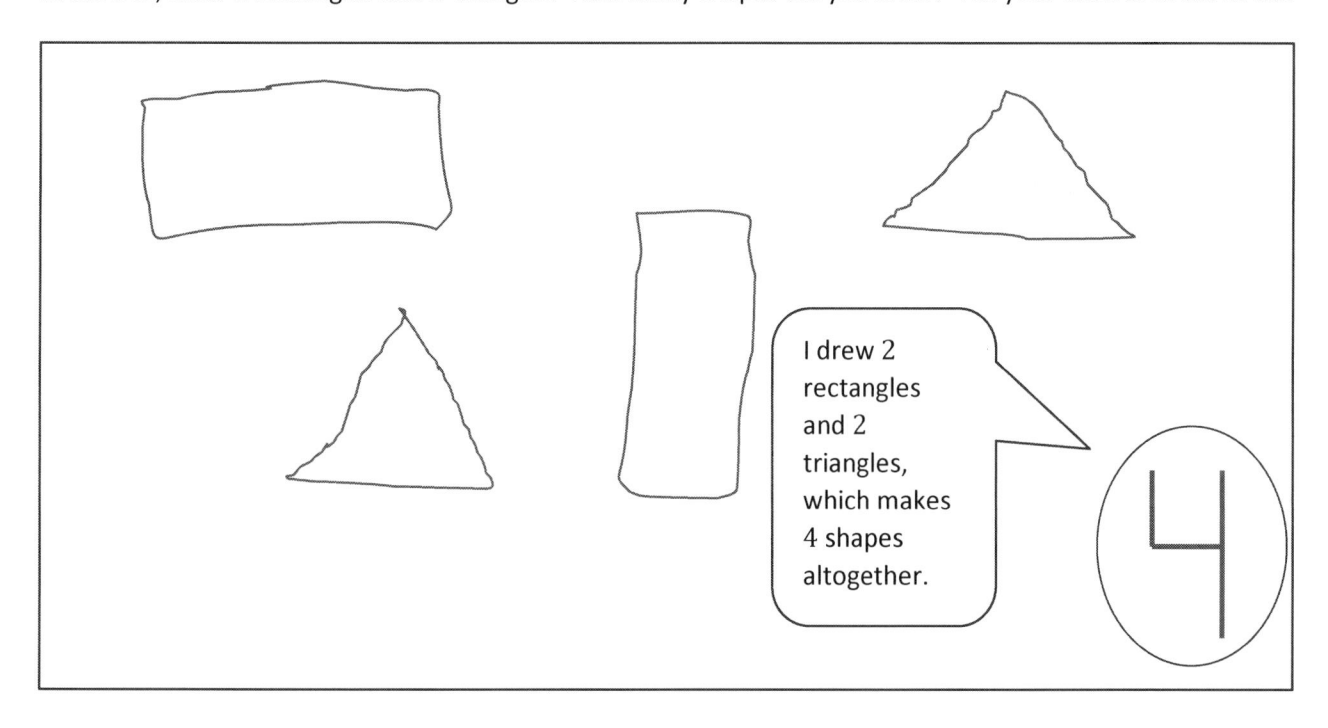

I drew 2 rectangles and 2 triangles, which makes 4 shapes altogether.

4

Lesson 3: Explain decisions about classifications of rectangles into categories using variants and non-examples. Identify shapes as rectangles.

GK-M2-Lesson 4

Color the triangles blue.
Color the rectangles red.
Color the circles green.
Color the hexagons yellow.

> This looks similar to a circle because it has curved lines and no corners. I know it is not a circle because it looks like it is stretched out.

> I know this isn't a shape because it is not closed.

> This shape does not look like a typical hexagon, but it has six straight sides, so it is a hexagon!

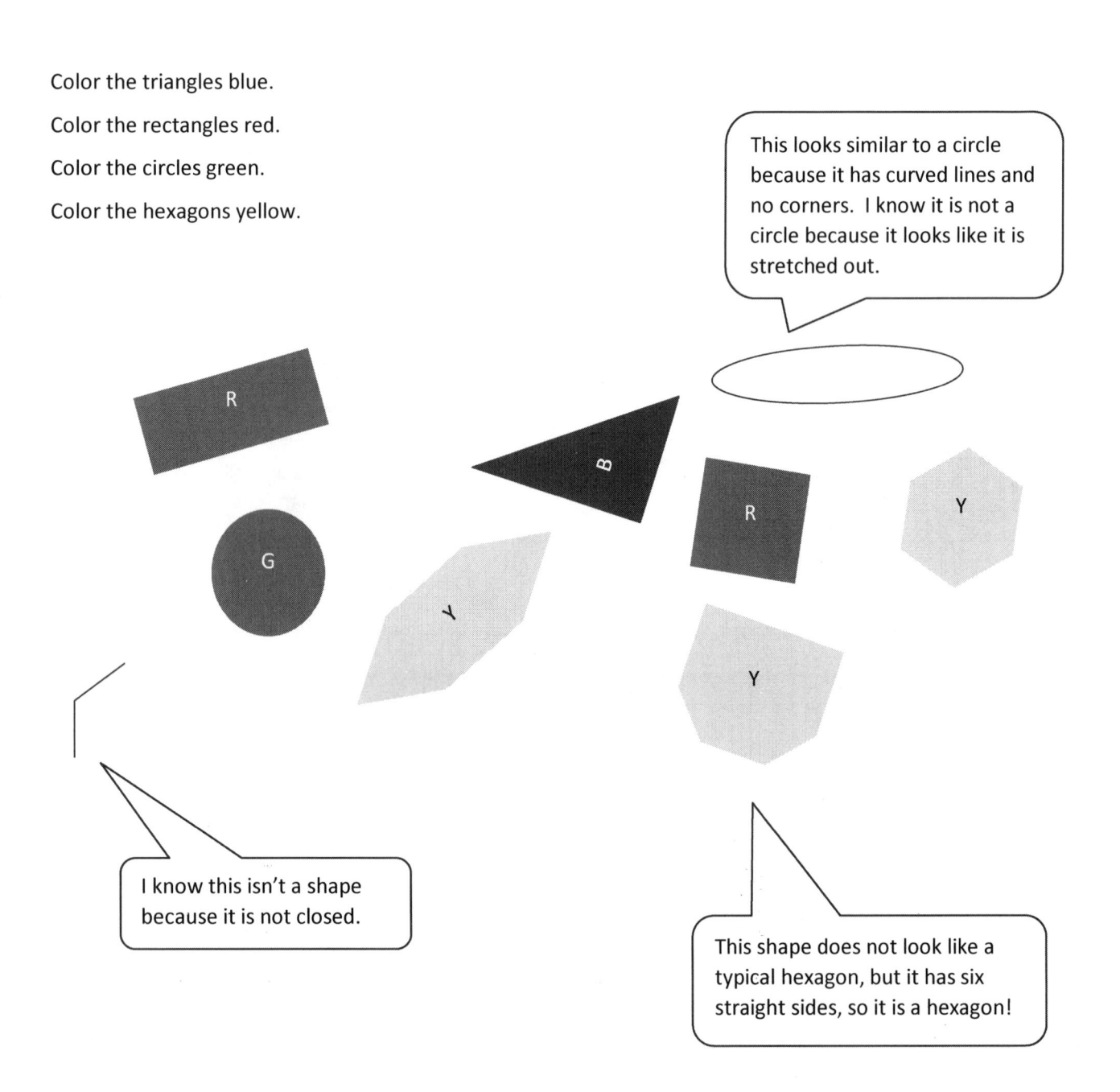

Homework Helper — A Story of Units K•2

GK-M2-Lesson 5

Next to the flower, draw a shape with 4 sides, 2 long and 2 short. Color it green.

Below the flower, draw a shape with no corners. Color it red.

Above the flower, draw a shape with 3 straight sides. Color it blue.

> I know that a circle has no corners. I drew the circle below the flower because below is the same as under.

In the box, draw 3 circles and 2 triangles. How many shapes did you draw? Put your answer in the circle.

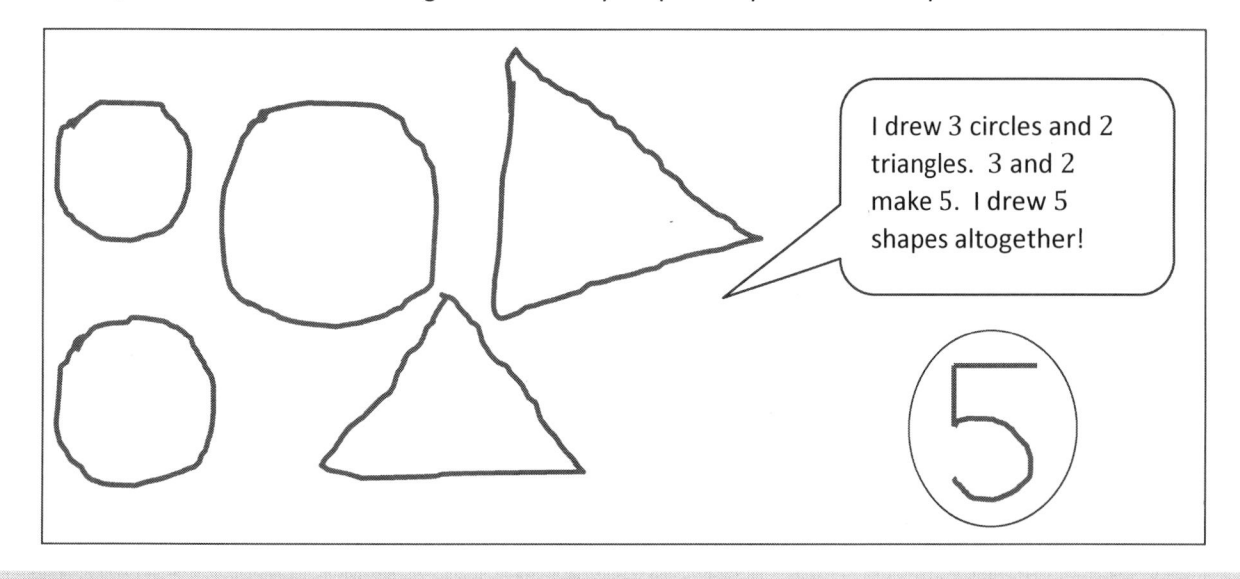

> I drew 3 circles and 2 triangles. 3 and 2 make 5. I drew 5 shapes altogether!

Lesson 5: Describe and communicate positions of all flat shapes using the words above, below, beside, in front of, next to, and behind.

Homework Helper — A Story of Units — K•2

GK-M2-Lesson 6

Find things in your house or in a magazine that look like these solids. Draw the solids or cut out and paste pictures from a magazine.

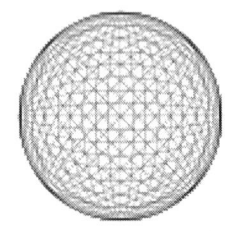

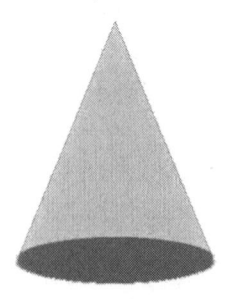

(Sphere shapes shown: golf ball and basketball)

(Cone shape shown alongside an ice cream cone)

> I know this shape! It is pointy at the end and holds ice cream!

(Cylinder shape shown alongside premium cans)

> These cans look the same as the solid shape because they both curve in the middle and have circles on the ends.

Lesson 6: Find and describe solid shapes using informal language without naming.

GK-M2-Lesson 7

Cut one set of solid shapes. Sort the 4 solid shapes. Paste them onto the chart.

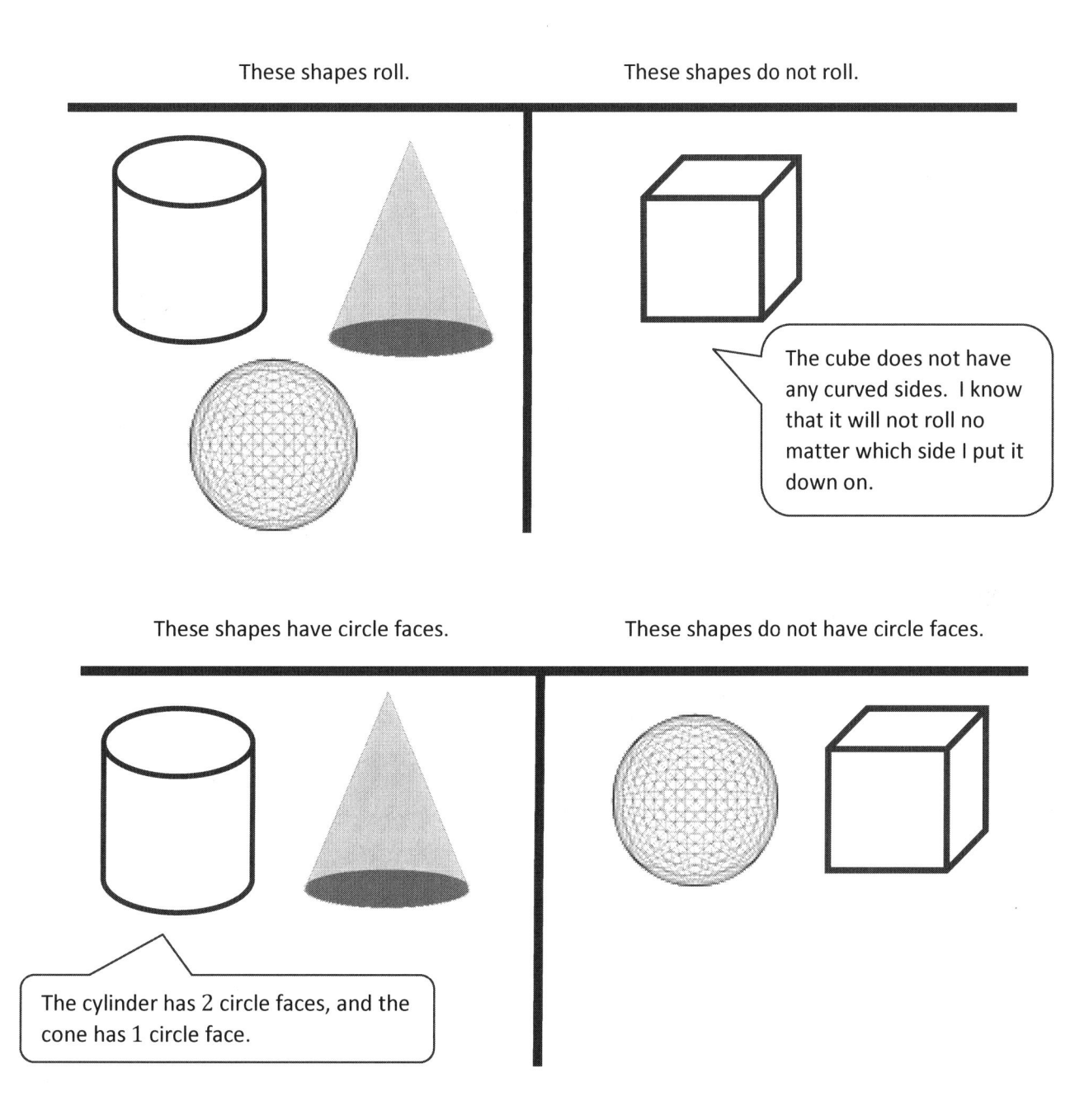

These shapes roll.	These shapes do not roll.
cylinder, cone, sphere	cube — The cube does not have any curved sides. I know that it will not roll no matter which side I put it down on.
These shapes have circle faces.	**These shapes do not have circle faces.**
cylinder, cone — The cylinder has 2 circle faces, and the cone has 1 circle face.	sphere, cube

Lesson 7: Explain decisions about classification of solid shapes into categories. Name the solid shapes.

Homework Helper   A Story of Units   K•2

GK-M2-Lesson 8

Tell someone at home the names of each solid shape.

> I know the cone is above the car, so I colored it orange. The opposite of above is below. I colored the cylinder green because it is below or under the car.

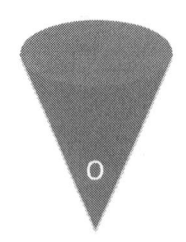

Cone

Sphere

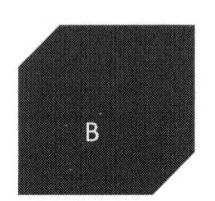

Cube

> Beside also means next to. I already colored the cube blue because it is in front of the car, so I knew the sphere was beside the car.

Cylinder

Color the shape in front of the car blue.

Color the shape above the car orange.

Color the shape below the car green.

Color the shape beside the car red.

Lesson 8: Describe and communicate positions of all solid shapes using the words above, below, beside, in front of, next to, and behind.

EUREKA MATH

Homework Helper   A Story of Units

GK-M2-Lesson 9

In each row, circle the one that doesn't belong. Explain your choice to a grown-up.

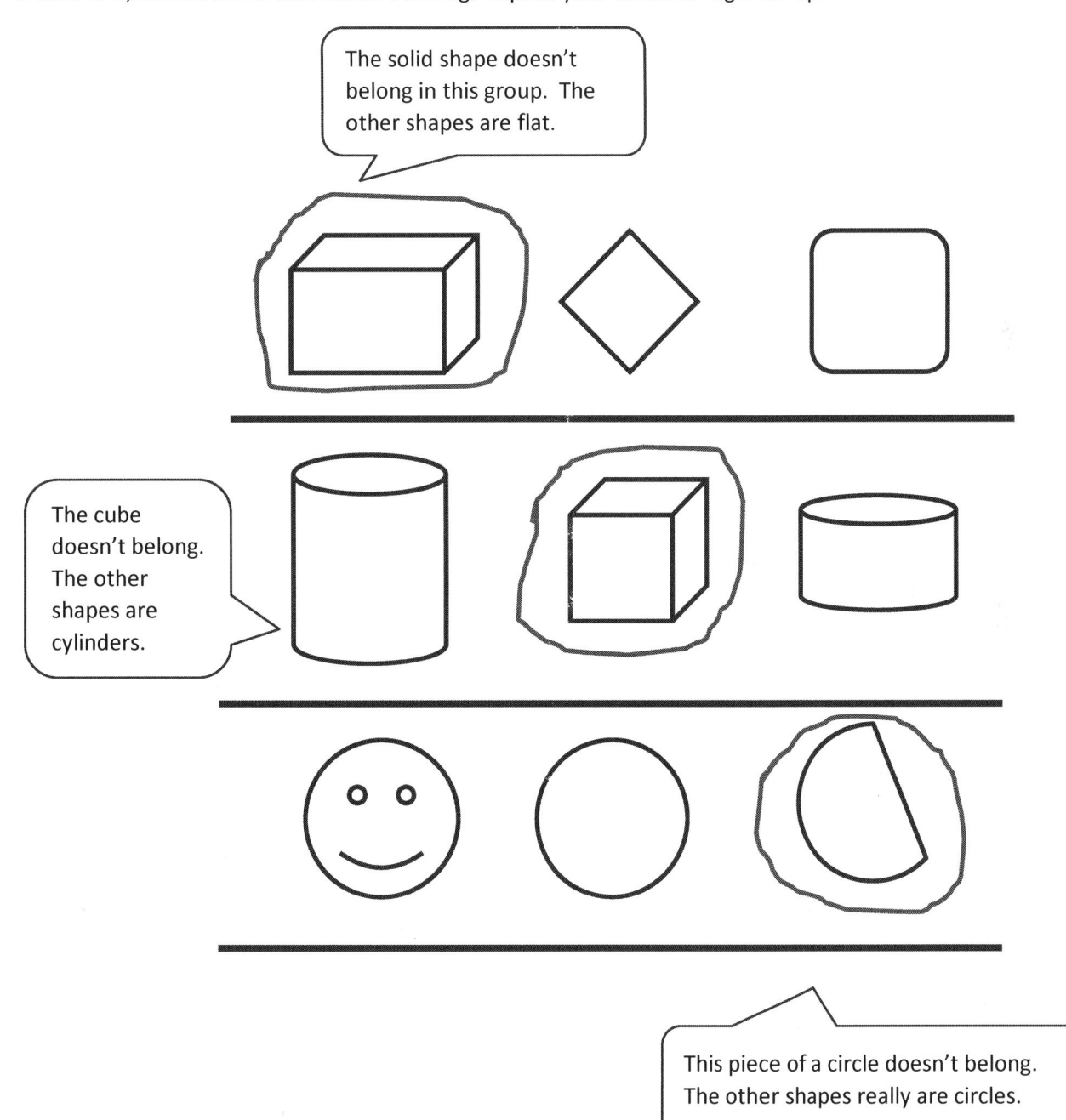

The solid shape doesn't belong in this group. The other shapes are flat.

The cube doesn't belong. The other shapes are cylinders.

This piece of a circle doesn't belong. The other shapes really are circles.

Lesson 9: Identify and sort shapes as two-dimensional or three-dimensional, and recognize two-dimensional and three-dimensional shapes in different orientations and sizes.

GK-M2-Lesson 10

Search your house to see what shapes and solids you can find. Draw the shapes that you see by tracing the faces of the solids that you find. Color your collage.

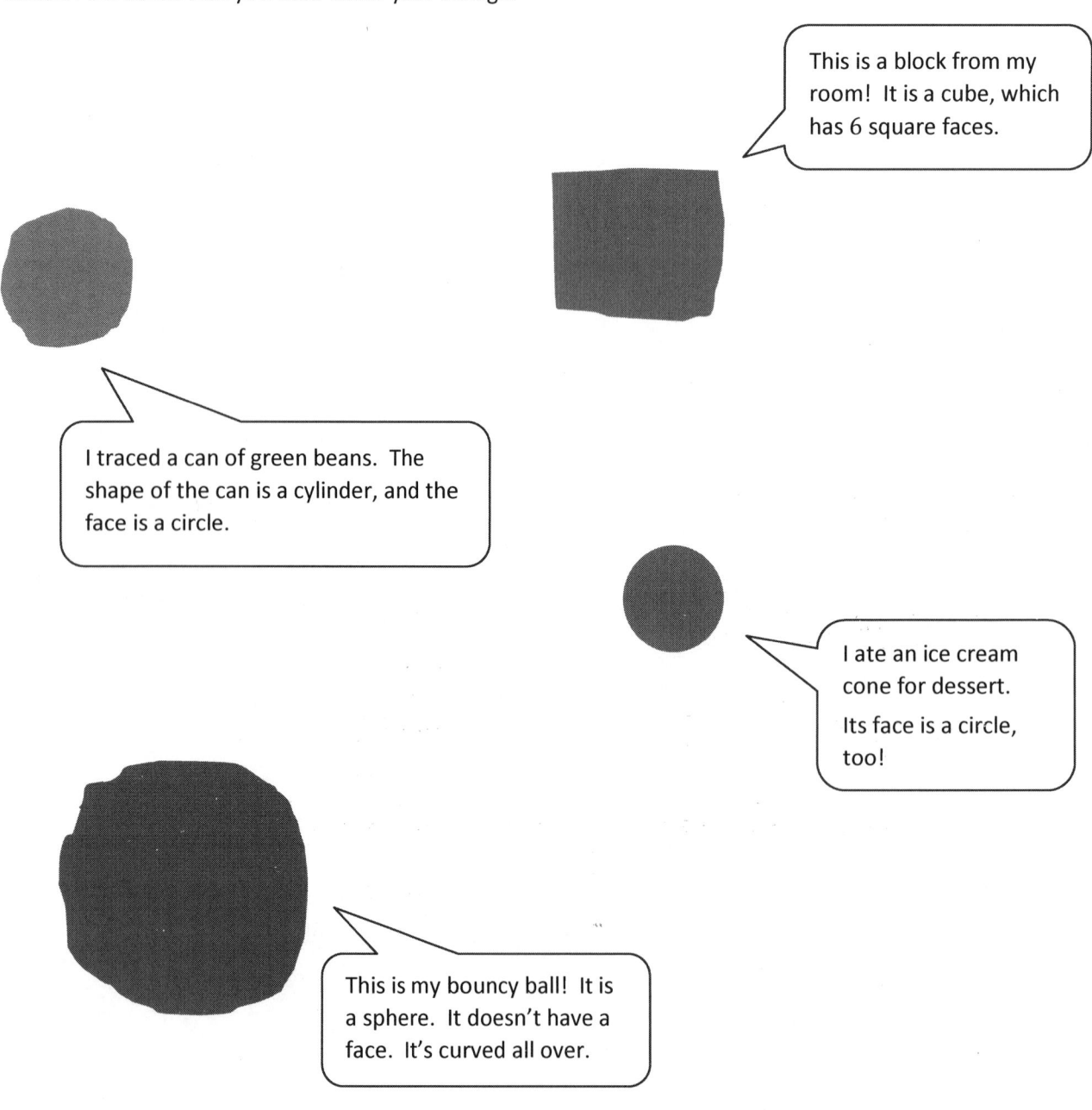

This is a block from my room! It is a cube, which has 6 square faces.

I traced a can of green beans. The shape of the can is a cylinder, and the face is a circle.

I ate an ice cream cone for dessert. Its face is a circle, too!

This is my bouncy ball! It is a sphere. It doesn't have a face. It's curved all over.

Homework Helpers

Grade K
Module 3

GK-M3-Lesson 1

Draw 2 more trees that are shorter than these trees.

Count how many trees you have now.

Write the number in the box.

$\boxed{4}$

On the back of your paper, draw something that is shorter than the refrigerator.

My kitty stands beside the refrigerator. The refrigerator is so tall! Kitty is much shorter than the refrigerator.

GK-M3-Lesson 2

Using the 1-foot piece of string from class, find three items at home that are shorter than your piece of string and three items that are longer than your piece of string. Draw a picture of those objects on the chart. Try to find at least one thing that is about the same length as your string, and draw a picture of it on the back.

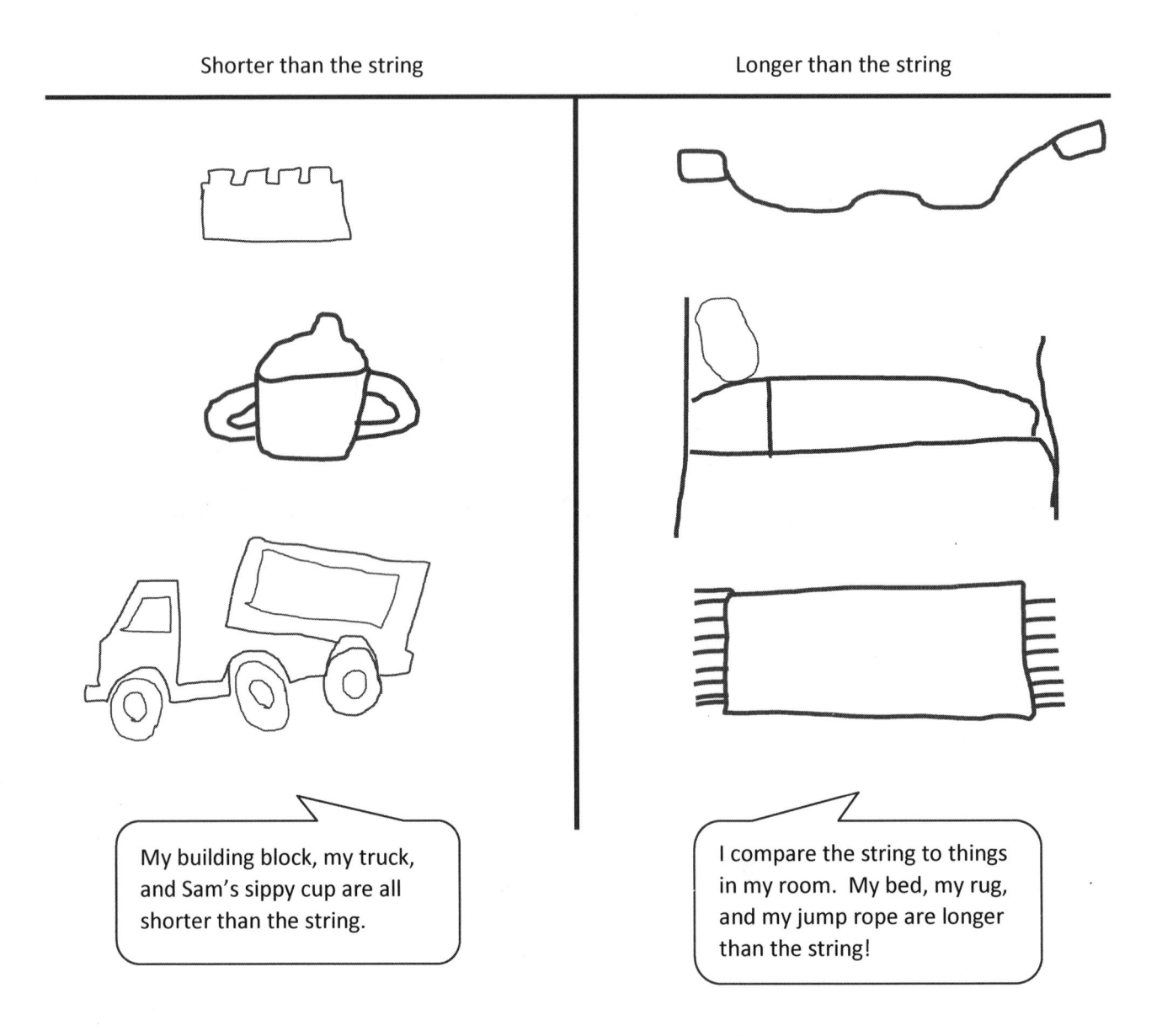

Shorter than the string	Longer than the string
My building block, my truck, and Sam's sippy cup are all shorter than the string.	I compare the string to things in my room. My bed, my rug, and my jump rope are longer than the string!

Lesson 2: Compare length measurements with string.

GK-M3-Lesson 3

Take out a new crayon. Use a red crayon to circle objects with lengths shorter than the crayon. Use a blue crayon to circle objects with lengths longer than the crayon.

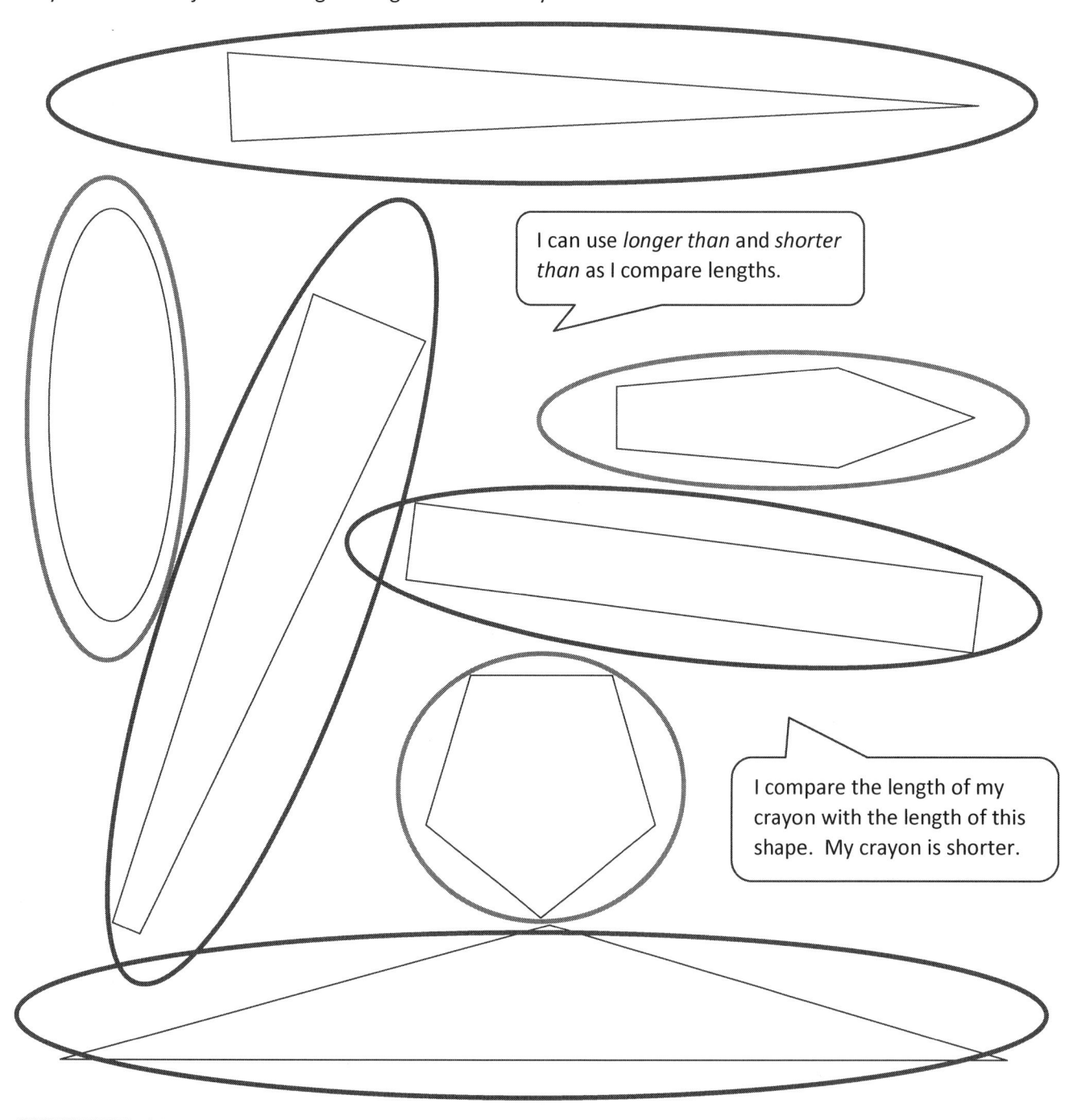

I can use *longer than* and *shorter than* as I compare lengths.

I compare the length of my crayon with the length of this shape. My crayon is shorter.

Lesson 3: Make a series of *longer than* and *shorter than* comparisons.

Homework Helper   A Story of Units   K•3

GK-M3-Lesson 4

Use a red crayon to circle the sticks that are shorter than the 5-stick.

> I count 5 cubes on this 5-stick.

> This stick has 3 cubes. It is shorter than the 5-stick.

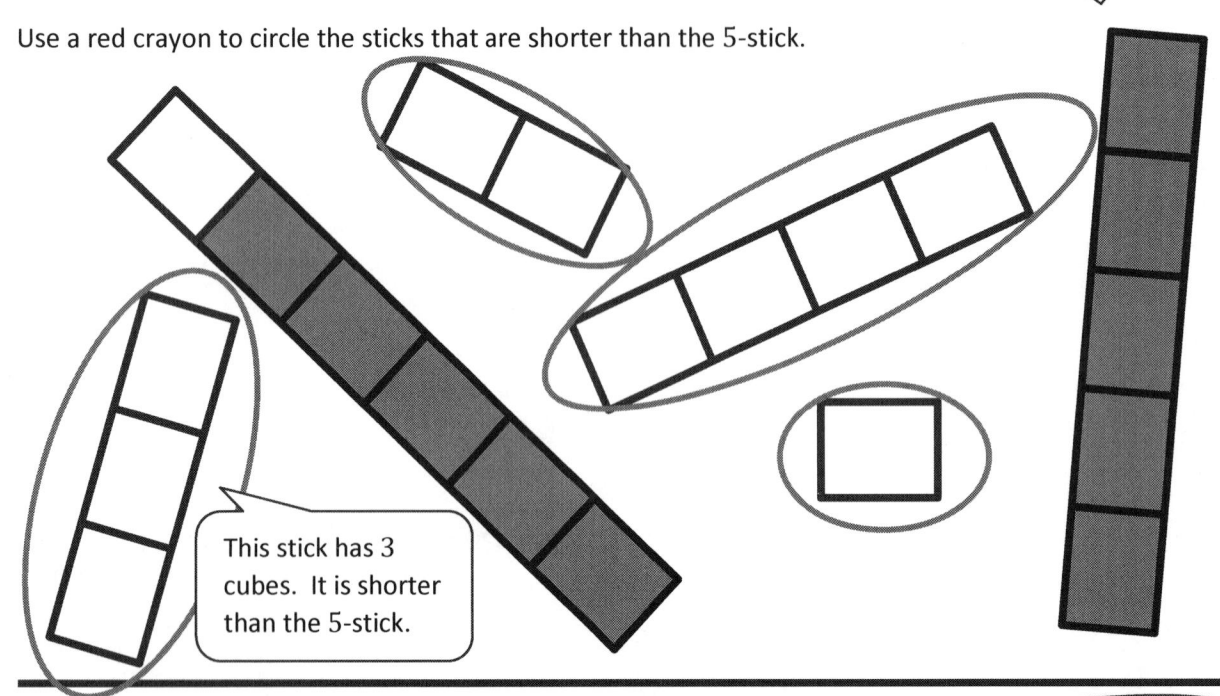

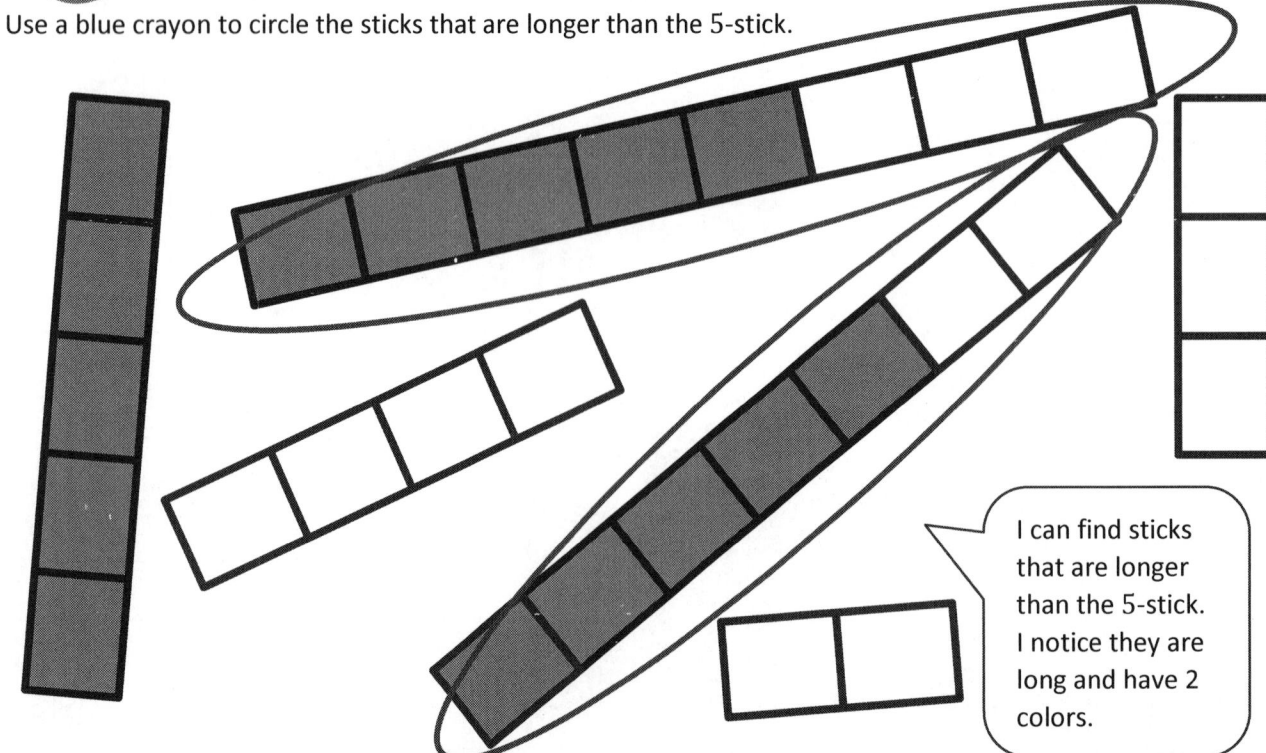

Use a blue crayon to circle the sticks that are longer than the 5-stick.

> I can find sticks that are longer than the 5-stick. I notice they are long and have 2 colors.

Lesson 4:   Compare the length of linking cube sticks to a 5-stick.

GK-M3-Lesson 5

Circle the stick that is shorter than the other.

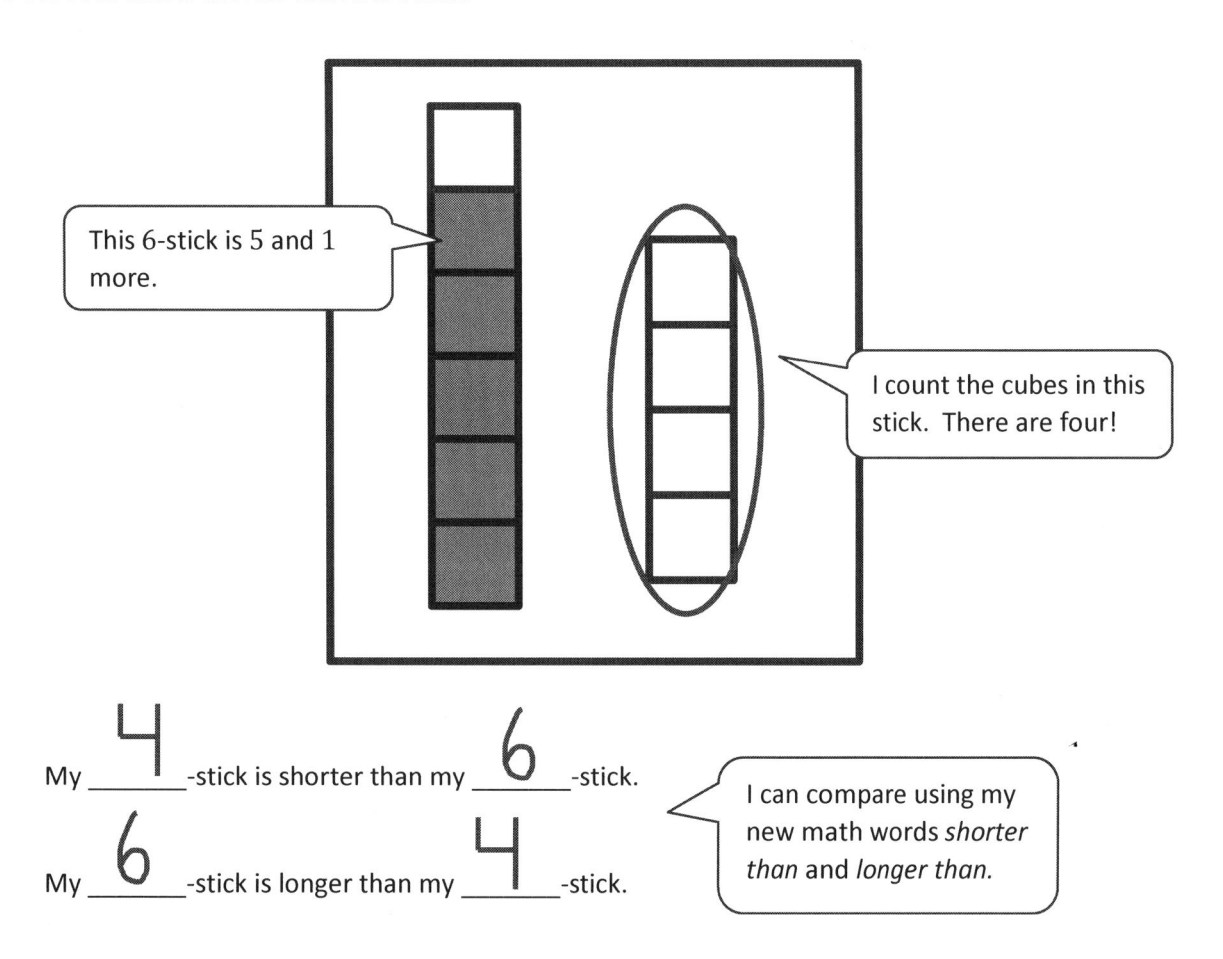

> This 6-stick is 5 and 1 more.

> I count the cubes in this stick. There are four!

My **4**-stick is shorter than my **6**-stick.

My **6**-stick is longer than my **4**-stick.

> I can compare using my new math words *shorter than* and *longer than*.

Draw a stick that is between a 3-stick and a 5-stick.

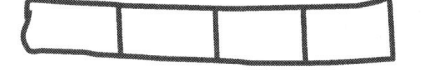

Draw a stick that is longer than your new stick.

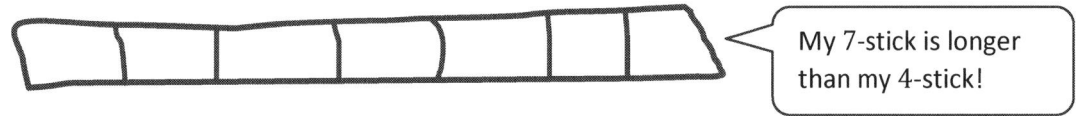

> My 7-stick is longer than my 4-stick!

Draw a stick that is shorter than your new stick.

Lesson 5: Determine which linking cube stick is *longer than* or *shorter than* the other.

Homework Helper

A Story of Units

K•3

GK-M3-Lesson 6

Color the cubes to show the length of the object.

The watermelon is shorter than the 8-stick.

I compare the piece of paper to the stick. The paper is about the same length as 4 cubes.

Lesson 6: Compare the length of linking cube sticks to various objects.

GK-M3-Lesson 7

These boxes represent cubes.

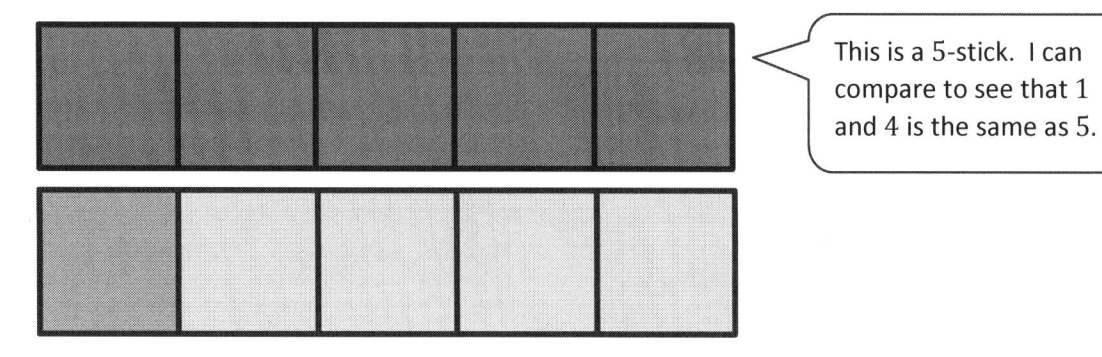

> This is a 5-stick. I can compare to see that 1 and 4 is the same as 5.

Color 1 cube green. Color 4 cubes red.
Together, my green 1-stick and red 4-stick are the same length as __5__ cubes.

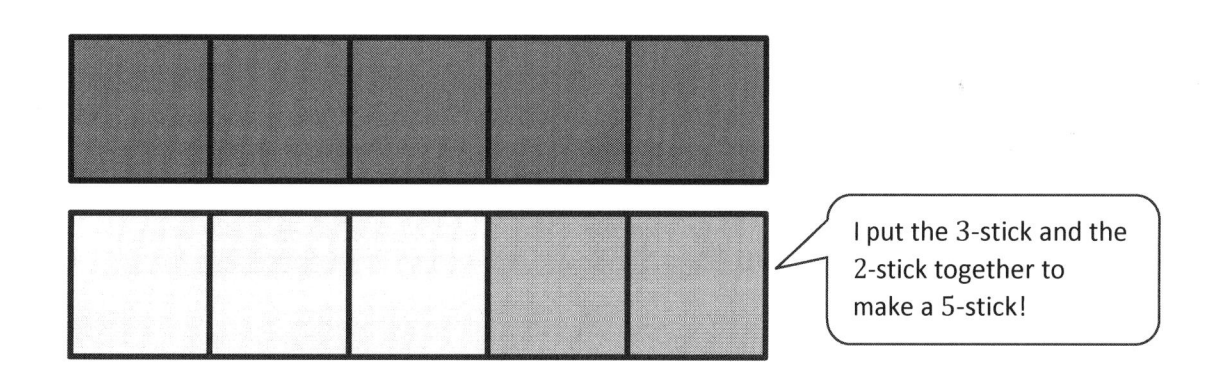

> I put the 3-stick and the 2-stick together to make a 5-stick!

Color 3 cubes yellow. Color 2 cubes blue.

How many cubes did you color? __5__
Together, my 3 cubes and 2 cubes are the same length as __5__ cubes.

Lesson 7: Compare objects using *the same as*.

Homework Helper   A Story of Units

GK-M3-Lesson 8

Draw an object that would be heavier than the one in the picture.

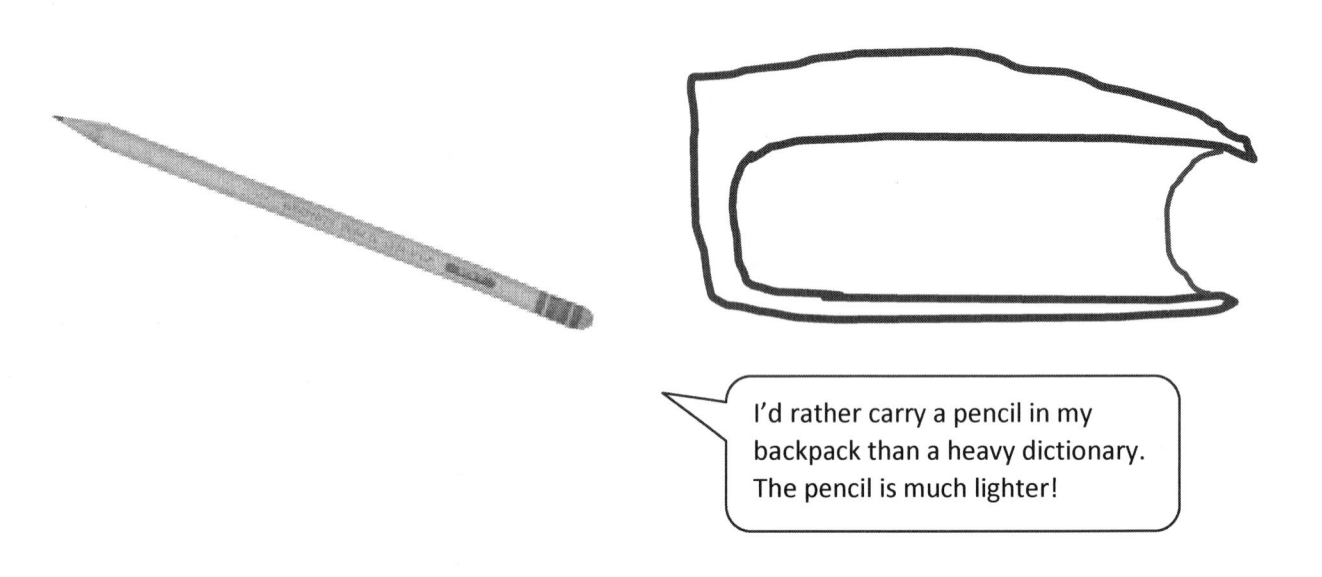

I'd rather carry a pencil in my backpack than a heavy dictionary. The pencil is much lighter!

It's easy for me to pick up the notebook. But, I can't pick up Daddy. He's too heavy.

Lesson 8:   Compare using *heavier than* and *lighter than* with classroom objects.

Homework Helper — A Story of Units

GK-M3-Lesson 9

Draw something inside the box that is heavier than the object on the balance.

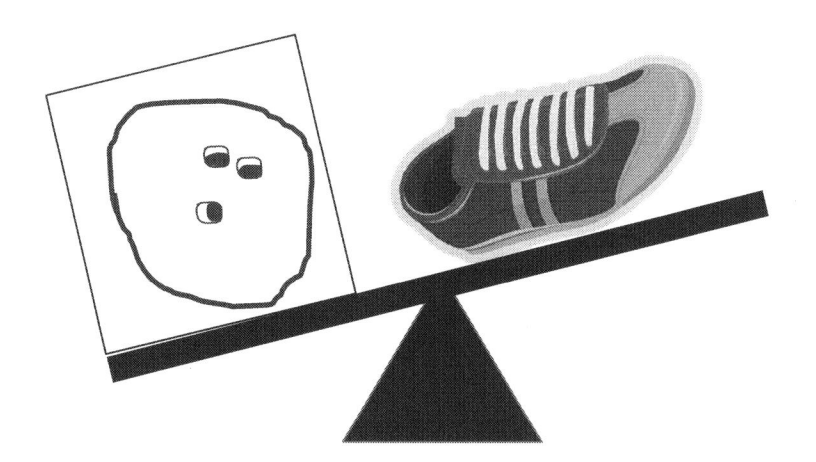

> A bowling ball is heavier than a shoe. It takes so many muscles to pick up a heavy bowling ball.

Draw something lighter than the object on the balance.

> An orange is lighter than a pineapple. It's easier to carry. It weighs so little, I can pick up more than one. I can even throw it!

Lesson 9: Compare objects using *heavier than, lighter than,* and *the same as* with balance scales.

Homework Helper

A Story of Units

GK-M3-Lesson 10

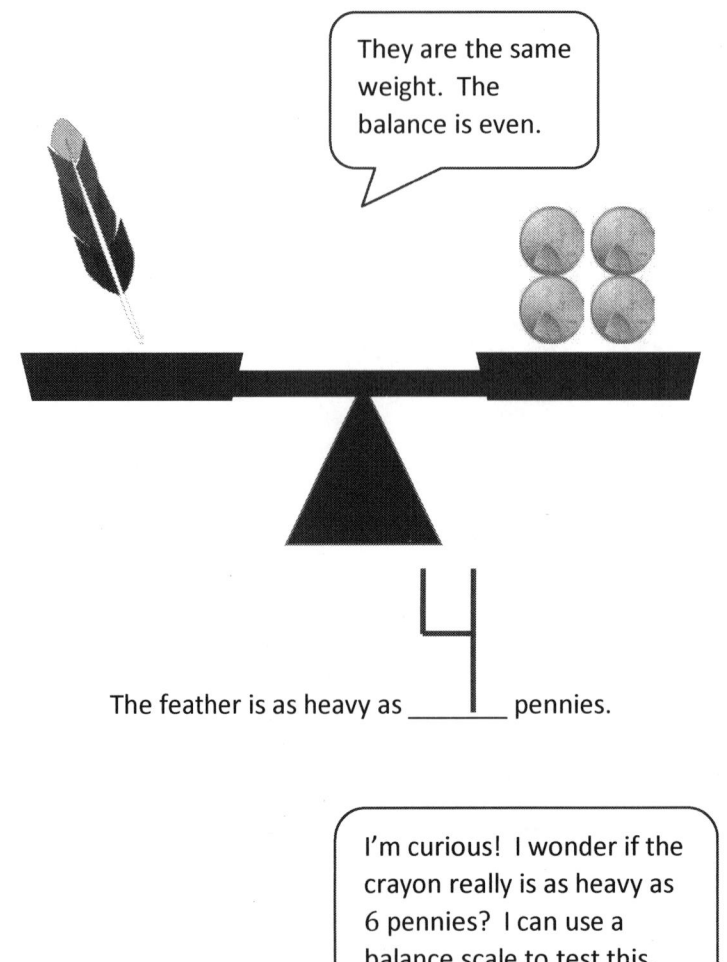

"They are the same weight. The balance is even."

The feather is as heavy as ____4____ pennies.

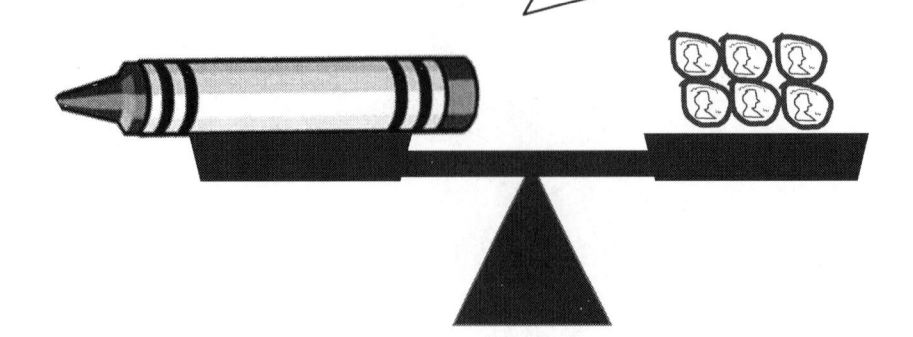

"I'm curious! I wonder if the crayon really is as heavy as 6 pennies? I can use a balance scale to test this."

Draw in the pennies so that the crayon is as heavy as 6 pennies.

Lesson 10: Compare the weight of an object to a set of unit weights on a balance scale.

GK-M3-Lesson 11

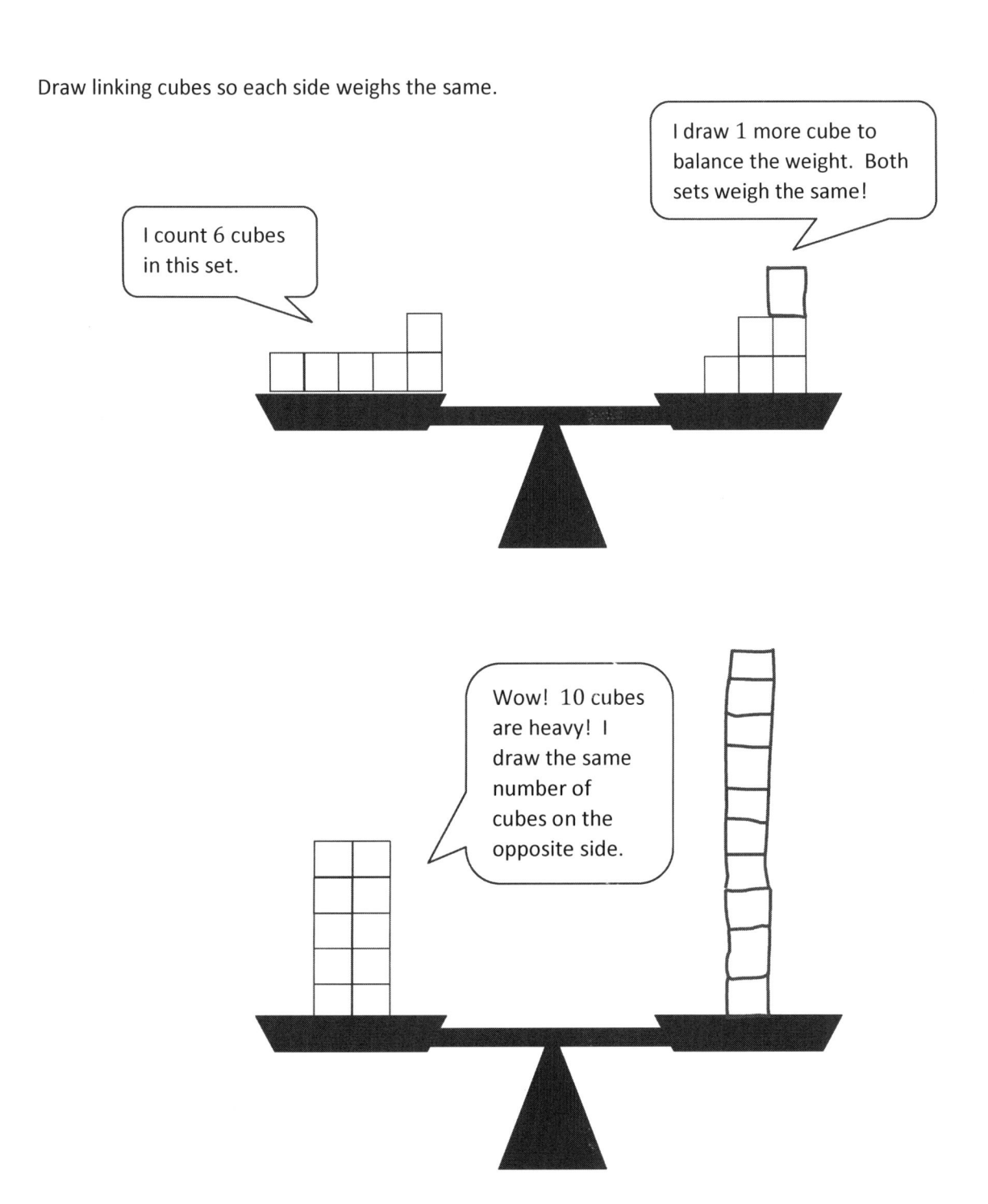

Draw linking cubes so each side weighs the same.

"I count 6 cubes in this set."

"I draw 1 more cube to balance the weight. Both sets weigh the same!"

"Wow! 10 cubes are heavy! I draw the same number of cubes on the opposite side."

Homework Helper — A Story of Units

GK-M3-Lesson 12

I can compare the weight of the apple with the weights of sets of different objects.

The apple is as heavy as __3__ cupcakes.

This is the same apple!

It takes more cupcakes to balance the weight of the apple than sneakers.

The apple is as heavy as __2__ sneakers.

Lesson 12: Compare the weight of an object with sets of different objects on a balance scale.

GK-M3-Lesson 13

Each rectangle shows 8 items. Circle two different sets within each. The two sets represent the two parts that make up the 8 objects.

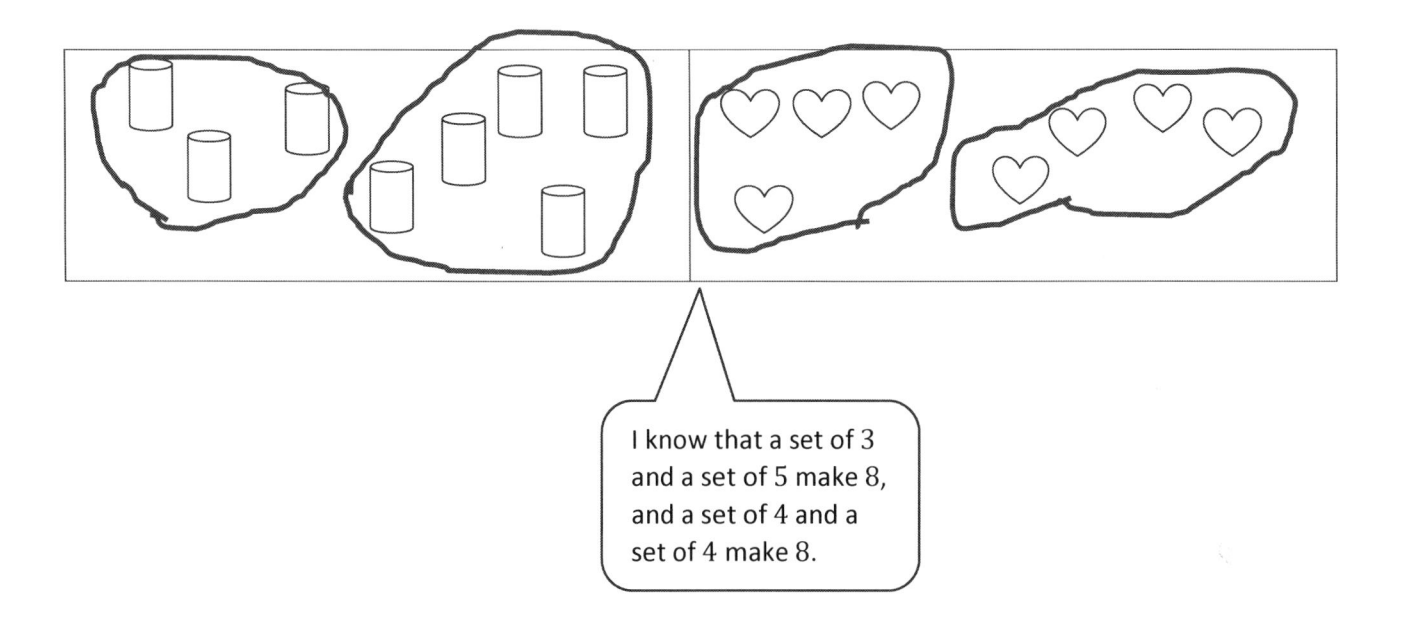

> I know that a set of 3 and a set of 5 make 8, and a set of 4 and a set of 4 make 8.

Lesson 13: Compare volume using *more than*, *less than* and *the same as* by pouring.

GK-M3-Lesson 14

Within each rectangle, make one set of 8 objects.

There are 9 clouds. I circle 8 of them. It's like 8 is hiding inside of 9.

Here I count 10, and I circle 8 of them.

GK-M3-Lesson 15

Circle 2 sets within each set of 8.

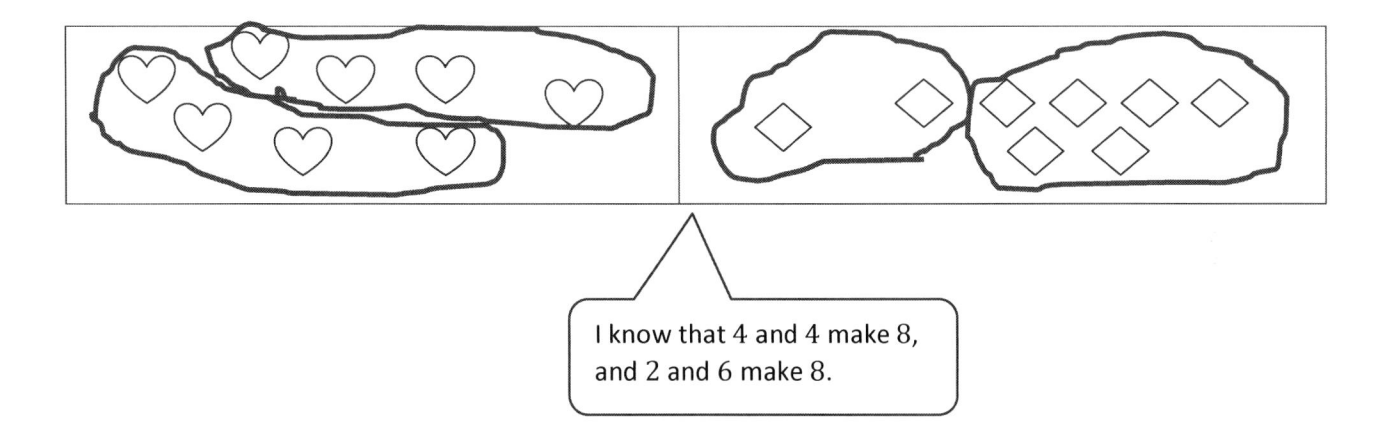

I know that 4 and 4 make 8, and 2 and 6 make 8.

Lesson 15: Compare using *the same as* with units.

GK-M3-Lesson 16

Cover the shape with beans. Count how many, and write the number in the box.

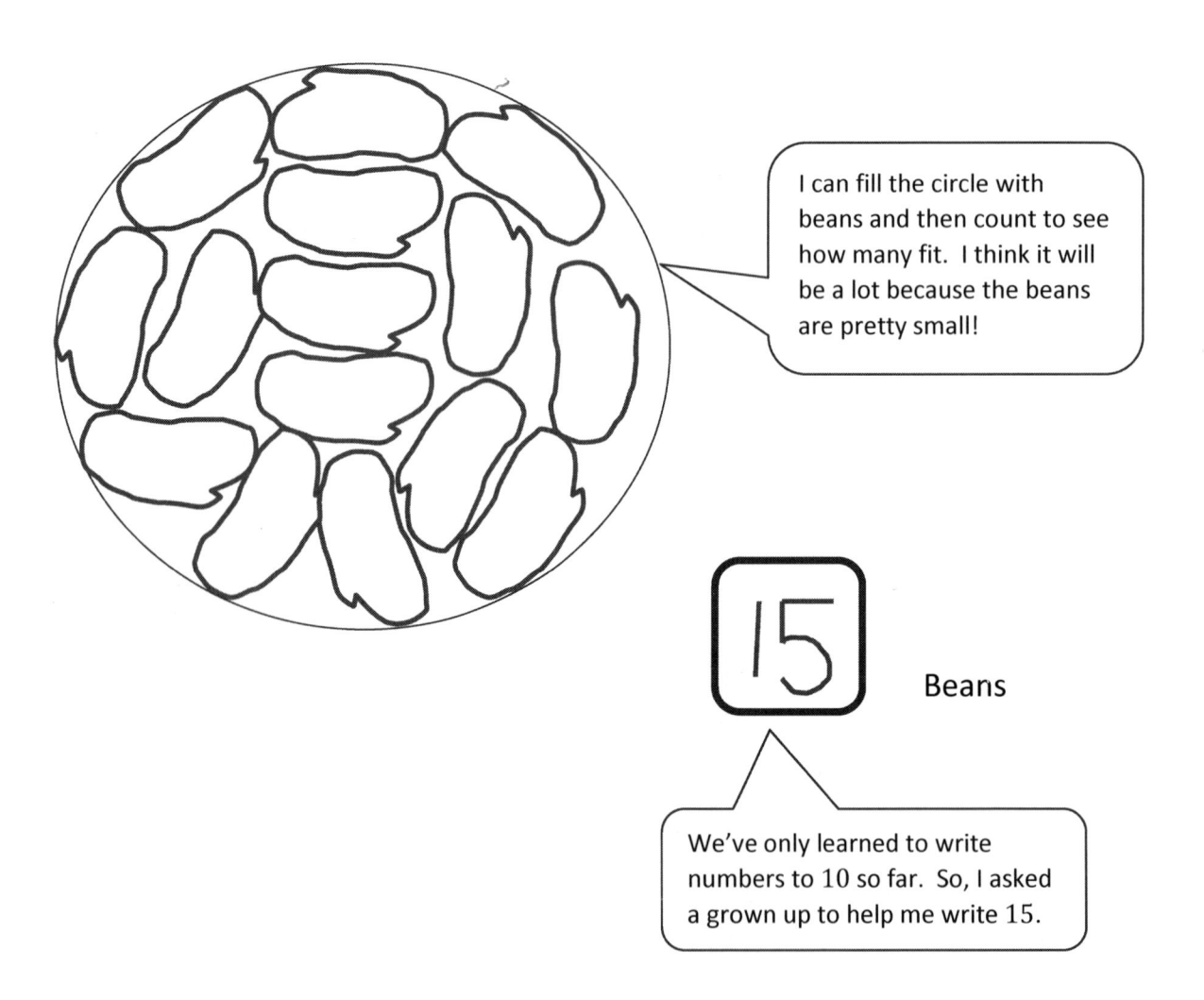

I can fill the circle with beans and then count to see how many fit. I think it will be a lot because the beans are pretty small!

15 Beans

We've only learned to write numbers to 10 so far. So, I asked a grown up to help me write 15.

Homework Helper   A Story of Units   K•3

GK-M3-Lesson 17

Draw straight lines with your ruler to see if there are enough flowers for the butterflies.

> I can draw a line to connect each butterfly with one flower.

> Then I keep going to see if there are enough flowers for every butterfly.

> Each butterfly gets one flower! That means there are enough!

Lesson 17:   Compare to find if there are enough.

You have 3 dog bones. Draw enough bowls so you can put 1 bone in each bowl.

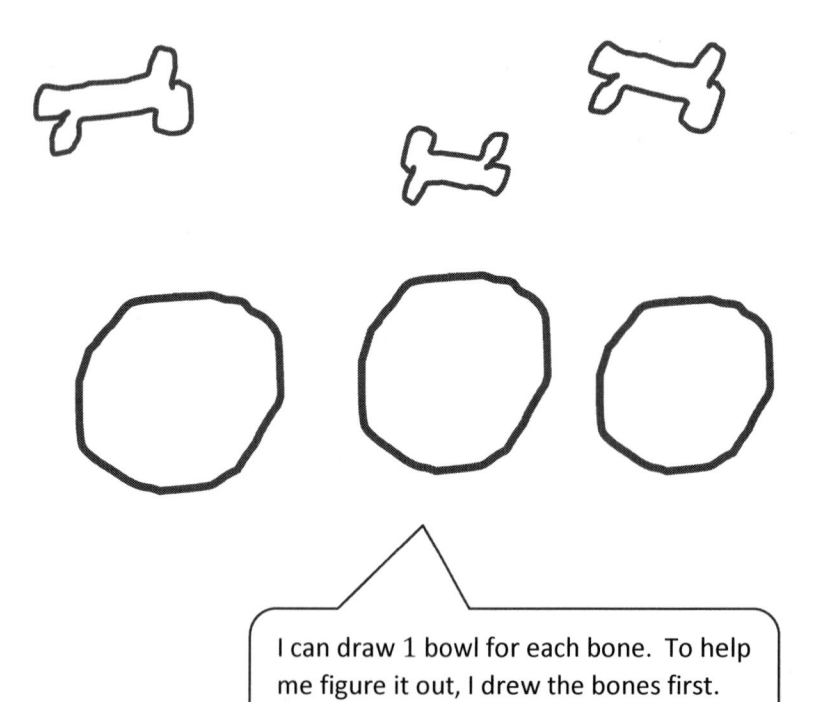

I can draw 1 bowl for each bone. To help me figure it out, I drew the bones first. There are 3 bones and 3 bowls.

Homework Helper

A Story of Units K•3

GK-M3-Lesson 18

Draw straight lines with your ruler to see if there are enough hats for the scarves.

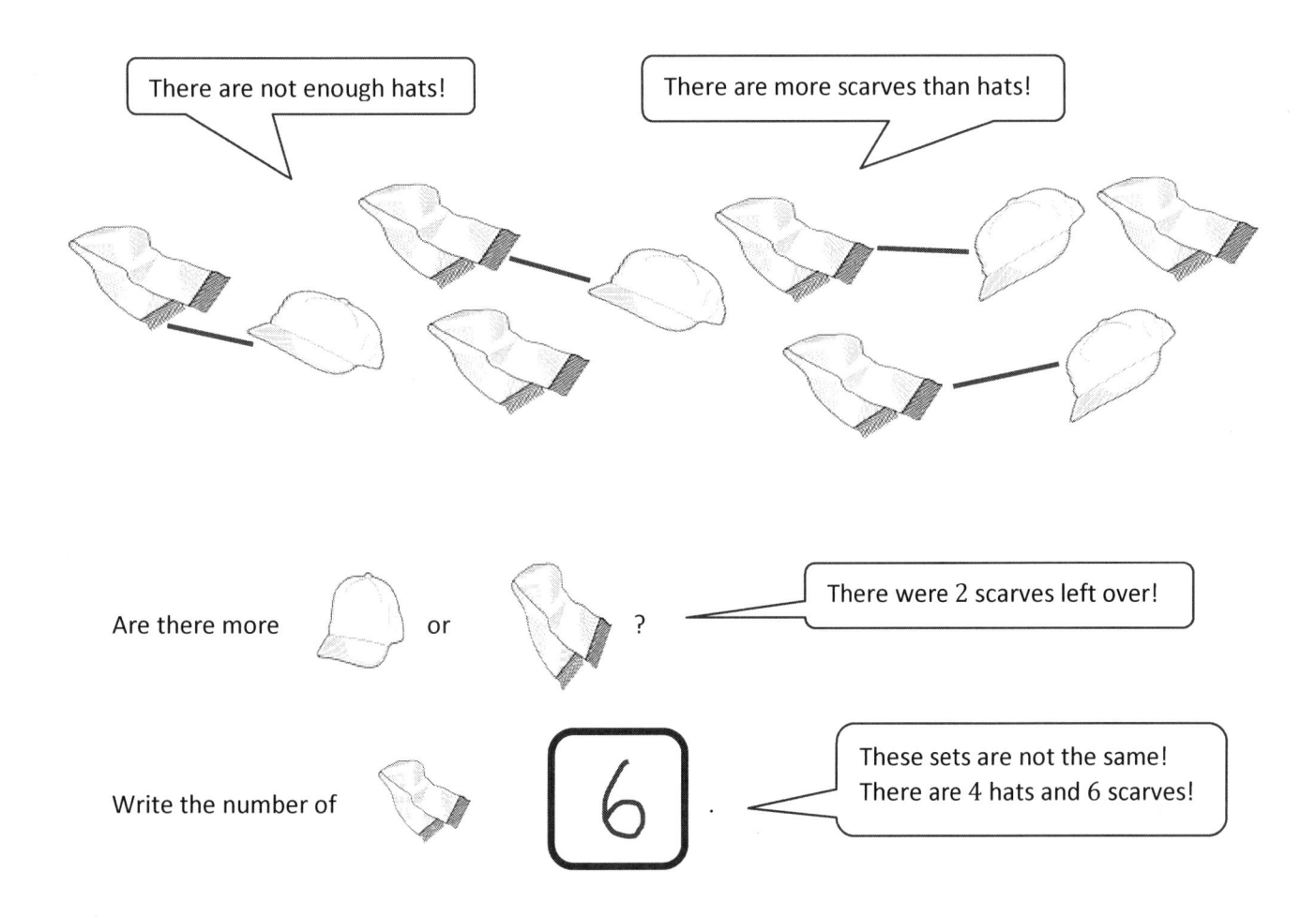

There are not enough hats!

There are more scarves than hats!

Are there more 🧢 or 🧣 ?

There were 2 scarves left over!

Write the number of 🧣 **6**

These sets are not the same! There are 4 hats and 6 scarves!

Lesson 18: Compare using *more than* and *the same as*.

Homework Helper

Write the number of  🍎. **6**

Write the number of 🐜. **6**

Are there the same number of 🍎 as 🐜? Circle **Yes** or No.

There are enough apples for each ant!

First I counted 6 apples. Then I counted 6 ants. These sets are the same!

Lesson 18: Compare using *more than* and *the same as*.

GK-M3-Lesson 19

Draw another bug so there are the same number of bugs as leaves.

> I can draw one more bug so there are the same number of bugs as leaves!

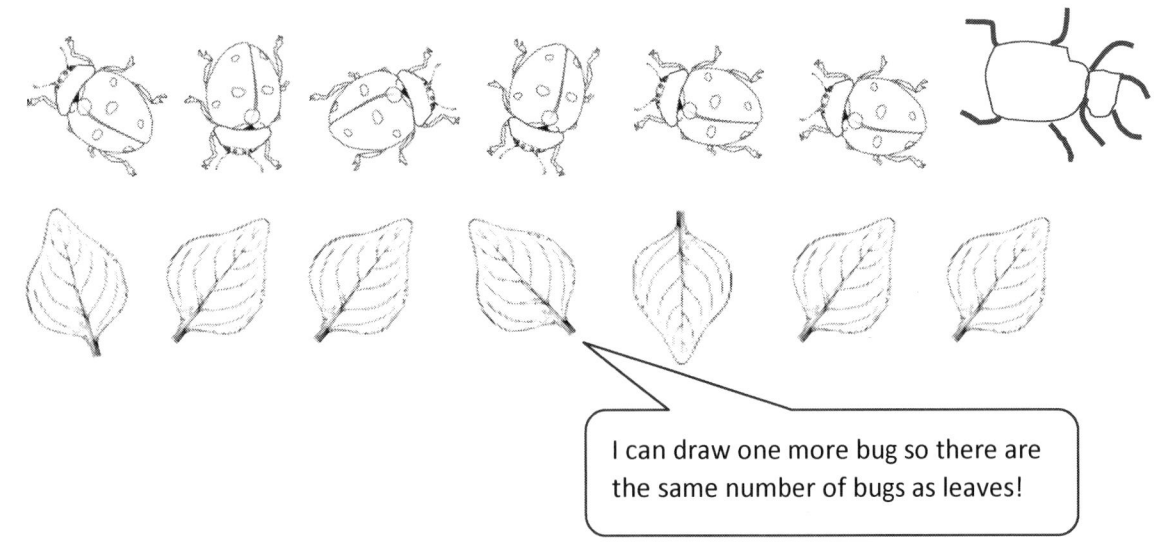

In the box below, draw 6 hearts.

Draw triangles so there are *fewer* triangles than hearts.

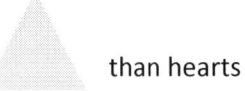

Draw circles so there are the *same* number of circles as hearts.

> I know I have fewer triangles than hearts because I drew 6 hearts, and I only drew 3 triangles!

Lesson 19: Compare using *fewer than* and *the same as*.

Homework Helper   A Story of Units   K•3

GK-M3-Lesson 20

On the first chain, color the first 4 beads orange.

On the next chain, color more than 4 beads purple.

How many beads did you color purple? Write the number in the box.

> I can make a row with more than 4! First I make a row the same size, and then I just color some more to make it longer.

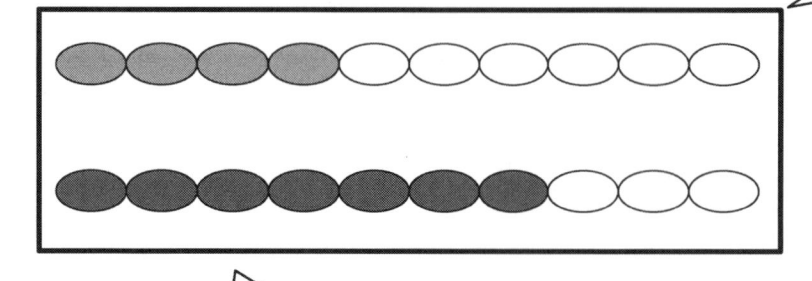

| 7 |

__7__ purple beads is more than 4.

> I know that 7 is more than 4 because the purple row of beads is longer!

Draw a chain with more than 5 beads but fewer than 9 beads.

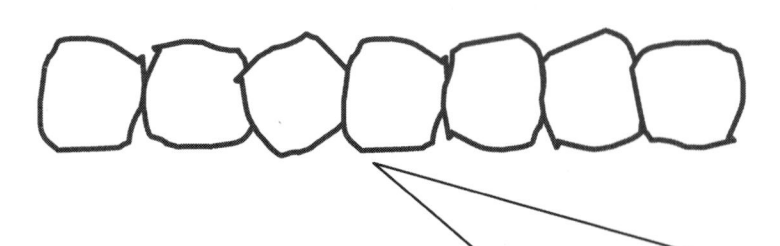

> I start by making 5 beads, and then I add more, one at a time. Before I get to 9, I stop. I stopped at 7. 7 is more than 5 but still fewer than 9.

Lesson 20: Relate *more* and *less* to length.

GK-M3-Lesson 21

Which has more? The 🦆 or 🐰

Circle the set that has more.

The set of ducks has more. I know because I count 6 ducks and only 4 bunnies. 6 is more than 4.

Draw a set of 3 kittens. Then draw some puppies. Are there fewer kittens or fewer puppies?

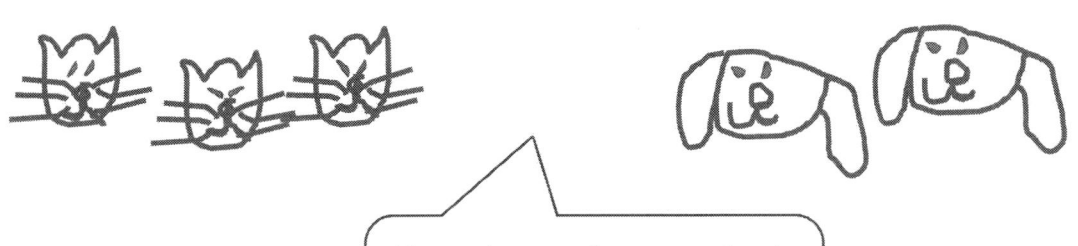

I know there are fewer puppies. I draw the 3 kittens, and then when I draw the puppies, I stop at 2.

Lesson 21: Compare sets informally using *more*, *less*, and *fewer*.

GK-M3-Lesson 22

Count the fish. In the next box, draw the same number of bowls as fish.

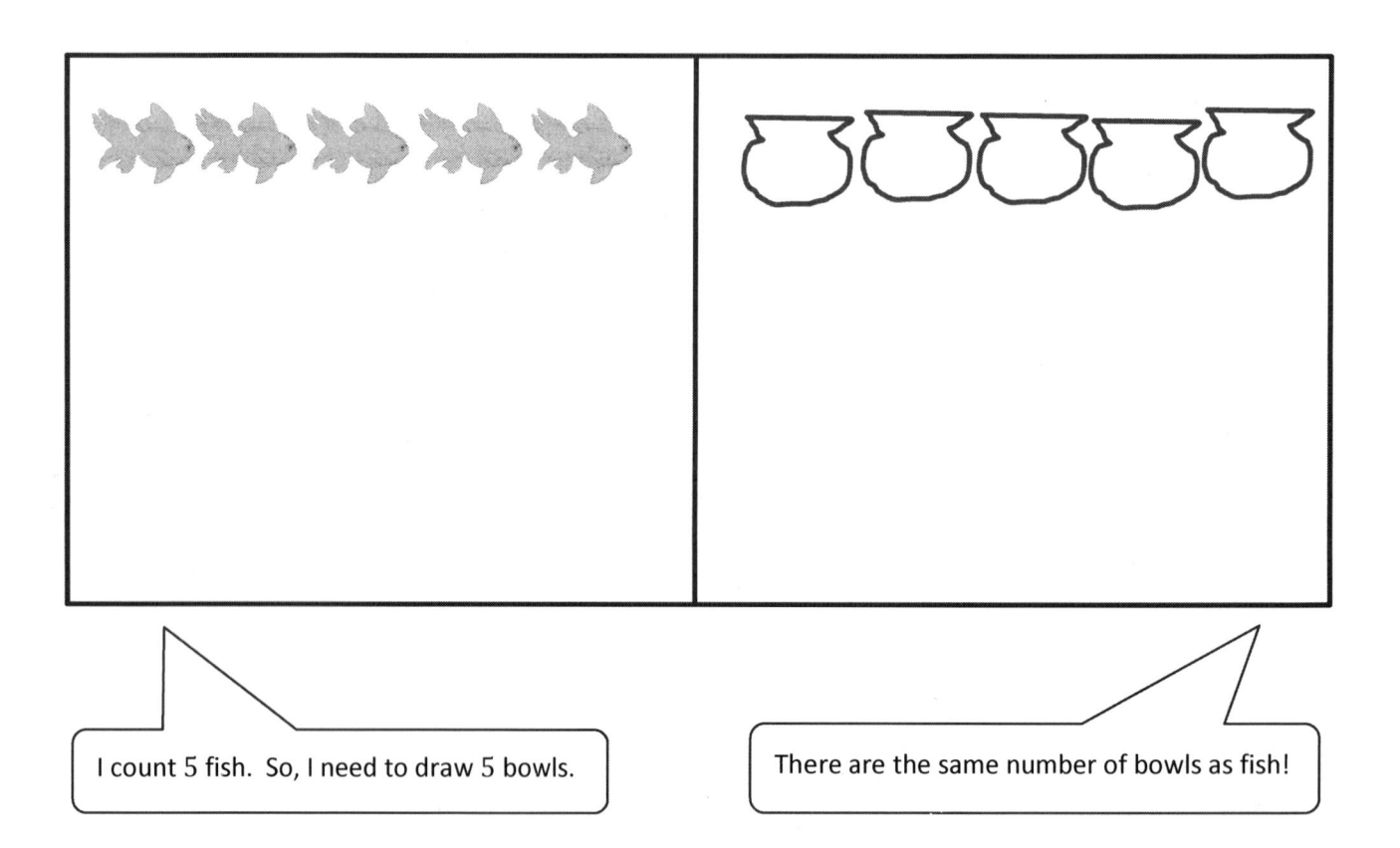

I count 5 fish. So, I need to draw 5 bowls.

There are the same number of bowls as fish!

GK-M3-Lesson 23

There are 4 snails!

How many snails? 4

Draw a leaf for every snail and one more leaf.

How many leaves? 5

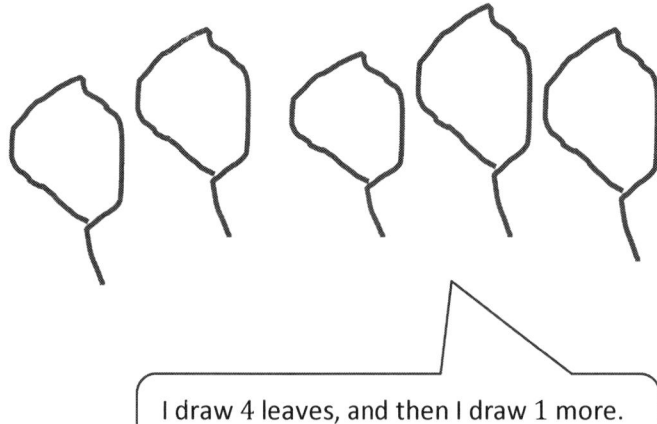

I draw 4 leaves, and then I draw 1 more. 1 more than 4 is 5.

Lesson 23: Reason to identify and make a set that has 1 more.

GK-M3-Lesson 24

Count the set of objects, and write how many in the box.

Draw a set of triangles that has 1 less, and write how many in the box. As you work, use your math words *less than*.

> I count 8 kites. Let me think. 8. 1 less is 7. So, I draw 7 triangles.

> 7 is less than 8.

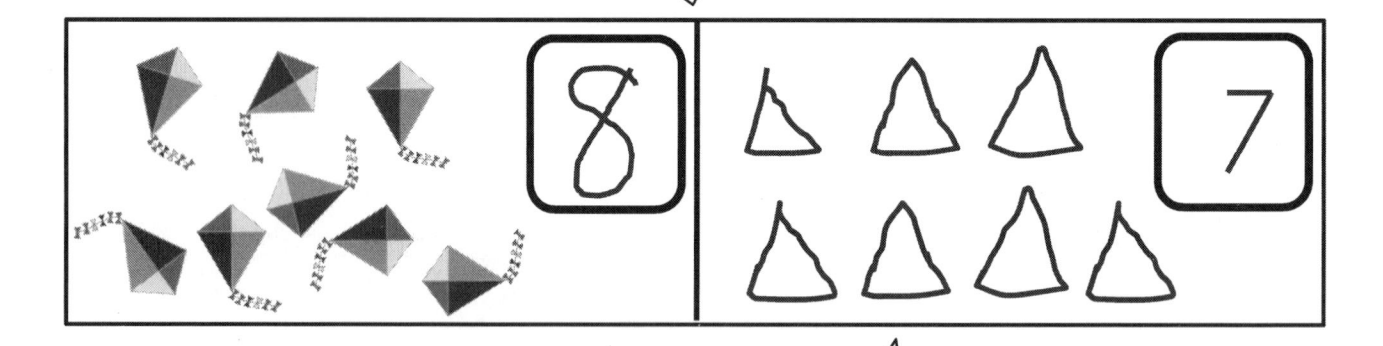

GK-M3-Lesson 25

Count the objects in each line. Write how many in the box. Then, fill in the blanks below. Use the words *more than* to compare the numbers.

__8__ is more than __7__.

I can see that there is 1 more cat! Then I counted 7 pandas and 8 cats.

8 is more than 7.

Lesson 25: Match and count to compare a number of objects. State which quantity is more.

Homework Helper — A Story of Units

GK-M3-Lesson 26

Count the objects in each line. Write how many in the box. Then, fill in the blanks below. Say your words *less than* out loud as you work.

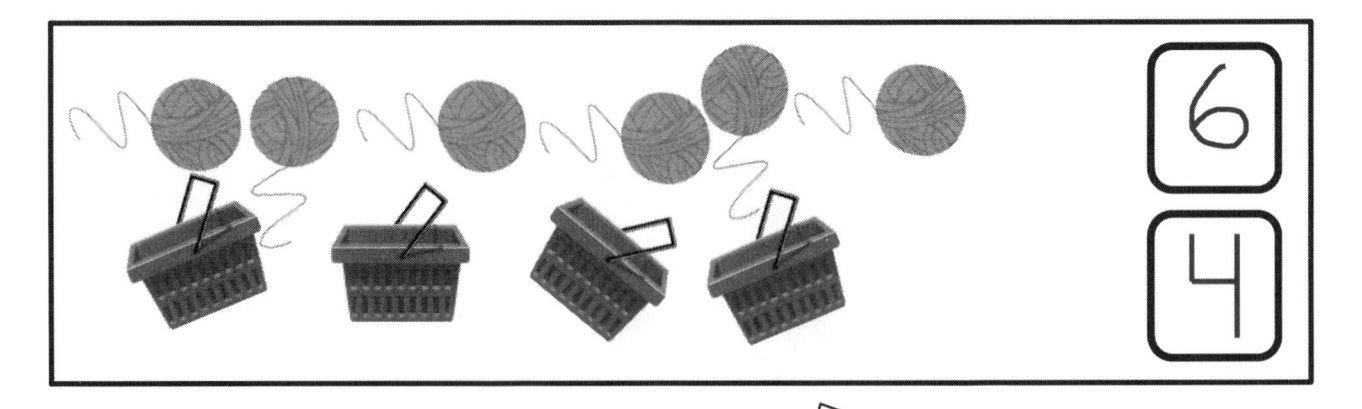

> 6
> 4

__4__ is less than __6__.

> There are not enough baskets for each ball of yarn to have a partner!

> 4 is less than 6. If I tried to put each ball of yarn into a basket, I would have some left over!

Lesson 26: Match and count to compare two sets of objects. State which quantity is less.

Homework Helper — A Story of Units

GK-M3-Lesson 27

Draw a tower with more cubes.

__4__ is more than __3__.

__3__ is less than __4__.

Draw a tower with fewer cubes.

__6__ is more than __3__.

__3__ is less than __6__.

> I can make a tower with more cubes. I just make it taller! The first tower has 3 cubes, so I made a tower with 1 more. My tower has 4 cubes.

> I can make a tower with fewer cubes. I just make it shorter! The first tower has 6 cubes, so I made mine with only 3 cubes. 3 is less than 6.

Lesson 27: Strategize to compare two sets.

GK-M3-Lesson 28

Visualize the number in Set A and Set B. Write the number in the sentences.

[6]   Set A

[3]   Set B

__6__ is more than __3__.

__3__ is less than __6__.

> I can see 6 in my head. 6 is more than 1 hand. 3 is less than 1 hand. 6 is more than 3.

GK-M3-Lesson 29

Draw a line from each container to the word that describes the amount of liquid the container is holding.

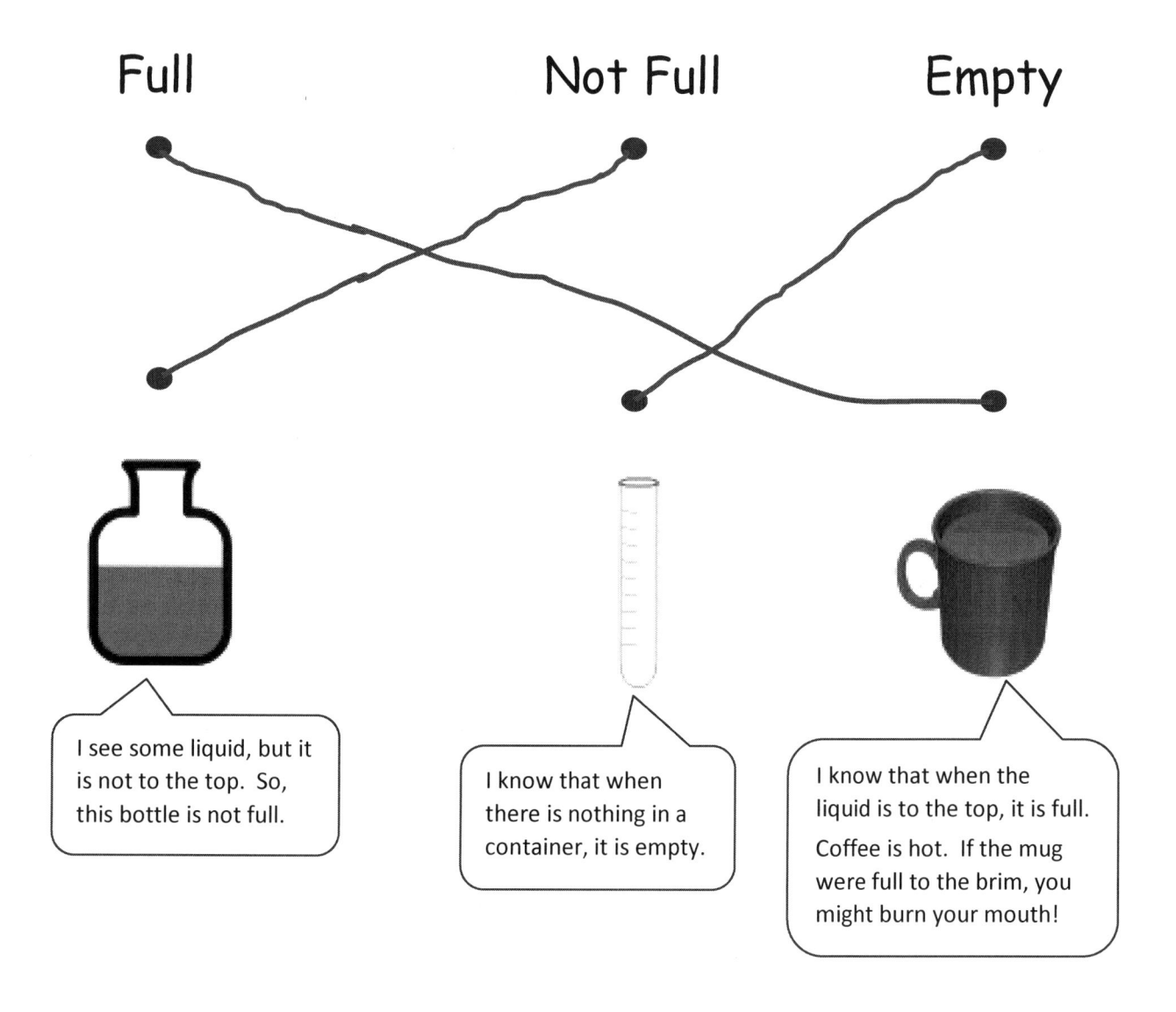

Full   Not Full   Empty

I see some liquid, but it is not to the top. So, this bottle is not full.

I know that when there is nothing in a container, it is empty.

I know that when the liquid is to the top, it is full.
Coffee is hot. If the mug were full to the brim, you might burn your mouth!

GK-M3-Lesson 30

Color 5 apples.

I colored __5__ apples.

I need to color __5__ more to make 10.

> There are 10 apples in all.
> I color 5 of them.
> I can count the rest to see how many more to make 10.

GK-M3-Lesson 31

Read the following directions to your child to make a house:

- Draw a square as wide as a fork.
- Draw a triangle on top of the square as tall as your pinky for the roof.
- Draw a rectangle as long as your thumb for the door.
- Draw 2 square windows each as long as a fingernail.

Hmm...let me look at my pinky to see how tall the roof should be.

Yes, that looks about right!

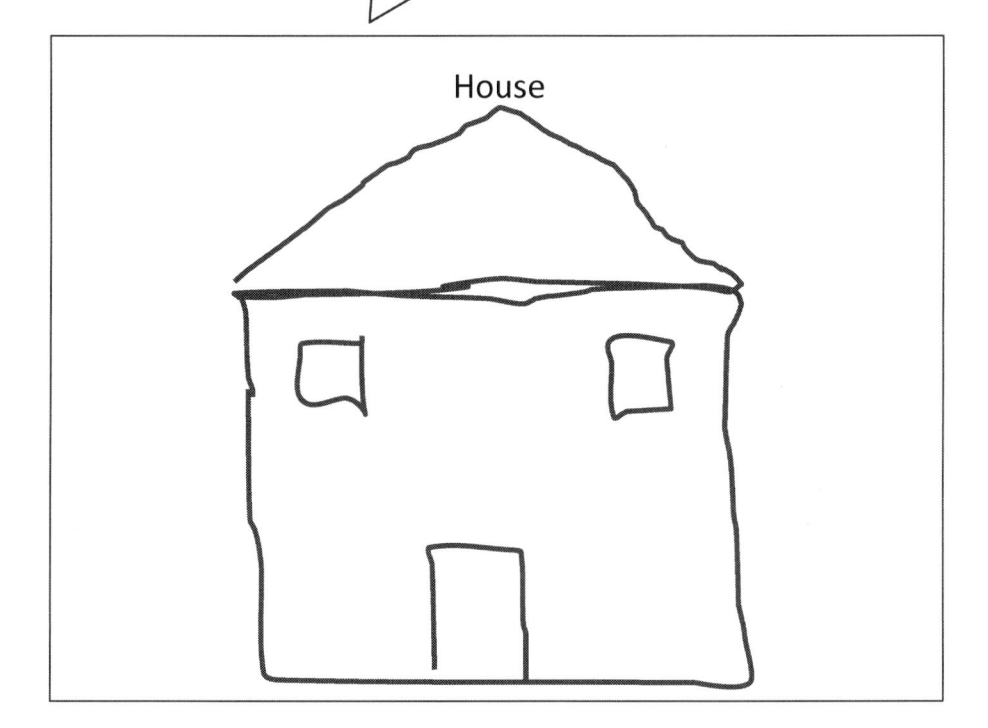

House

Lesson 31: Use benchmarks to create and compare rectangles of different lengths to make a city.

GK-M3-Lesson 32

Circle groups of dots. Then, fill in the blanks to make a number sentence.

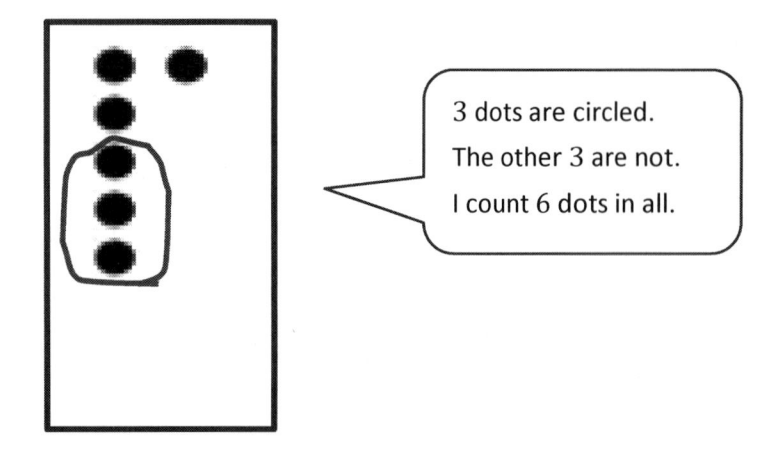

> 3 dots are circled.
> The other 3 are not.
> I count 6 dots in all.

3 and 3 is 6.

Make your own 6-dot card. Circle some dots, and then say, "____ and ____ is _____."

> 1 and 5 is 6.

Homework Helpers

Grade K
Module 4

GK-M4-Lesson 1

Number Bonds

Number bonds are models that show how numbers can be taken apart. The bigger number is the *whole*, or *total*, and the smaller numbers are the *parts* except when there is a 0. For now, please use everyday words such as "is," "and," and "make." Addition and subtraction will come later in this module. Number bonds are shown in different positions so that students can become flexible thinkers!

Draw the dark butterflies in the first circle on top. Draw the light butterflies in the next circle on top. Draw all the butterflies in the bottom circle.

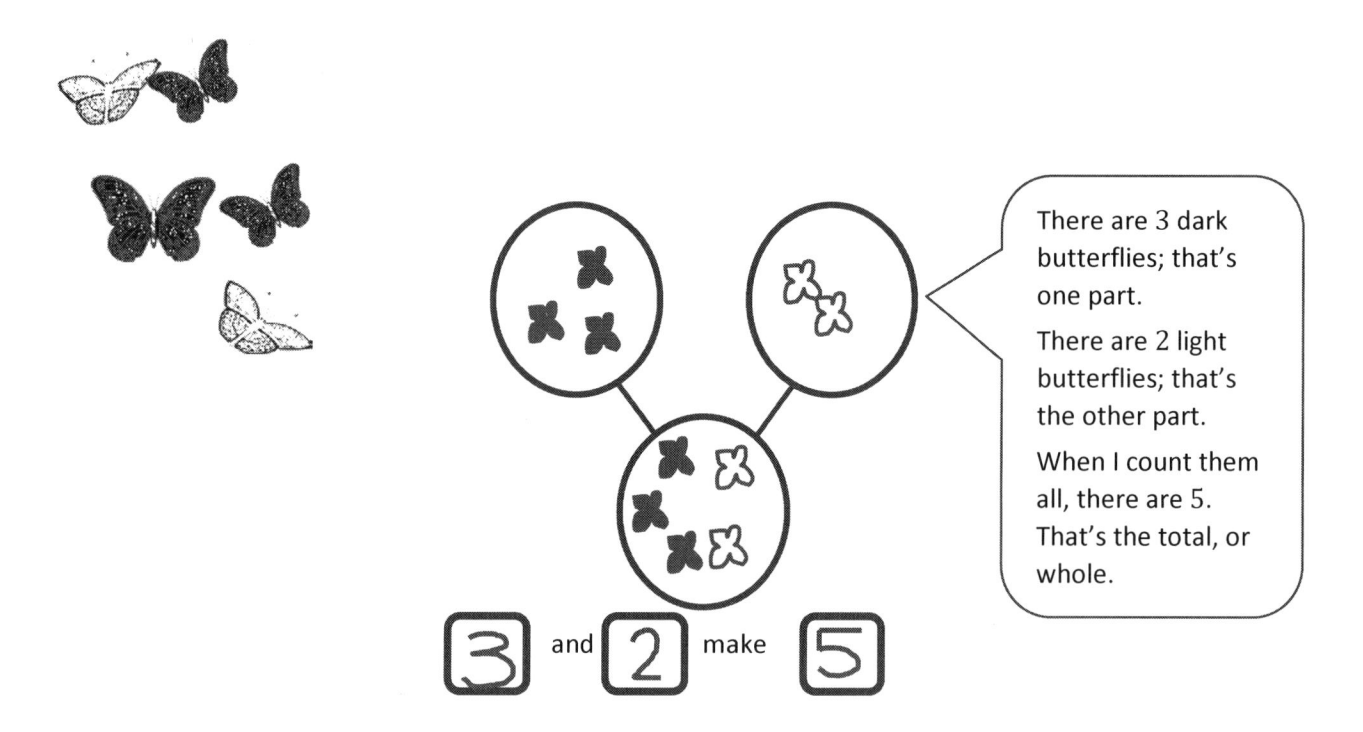

> There are 3 dark butterflies; that's one part.
>
> There are 2 light butterflies; that's the other part.
>
> When I count them all, there are 5. That's the total, or whole.

3 and 2 make 5

Lesson 1: Model composition and decomposition of numbers to 5 using actions, objects, and drawings.

Homework Helper   A Story of Units   K•4

GK-M4-Lesson 2

The squares below represent a cube stick. Color some squares blue and the rest of the squares red. Draw the squares you colored in the number bond. Show the hidden partners on your fingers to an adult. Color the fingers you showed.

> I decided to color 4 squares blue and 1 red. I could have colored 3 and 2. Any way I color, there are 5 squares in all.

> I show 4 fingers on one hand and 1 on the other hand. That's 5 fingers in all.
>
> Here are the fingers I showed. Can you think of another way?

> I see how my fingers, squares, and number bond match: 4 and 1 make 5. I can also say 5 is the same as 4 and 1.

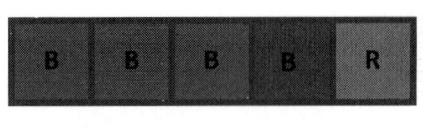

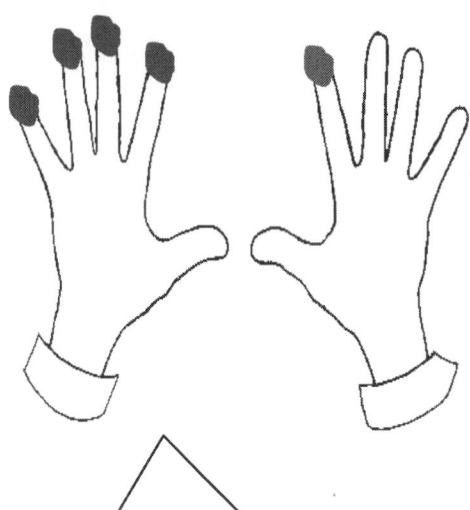

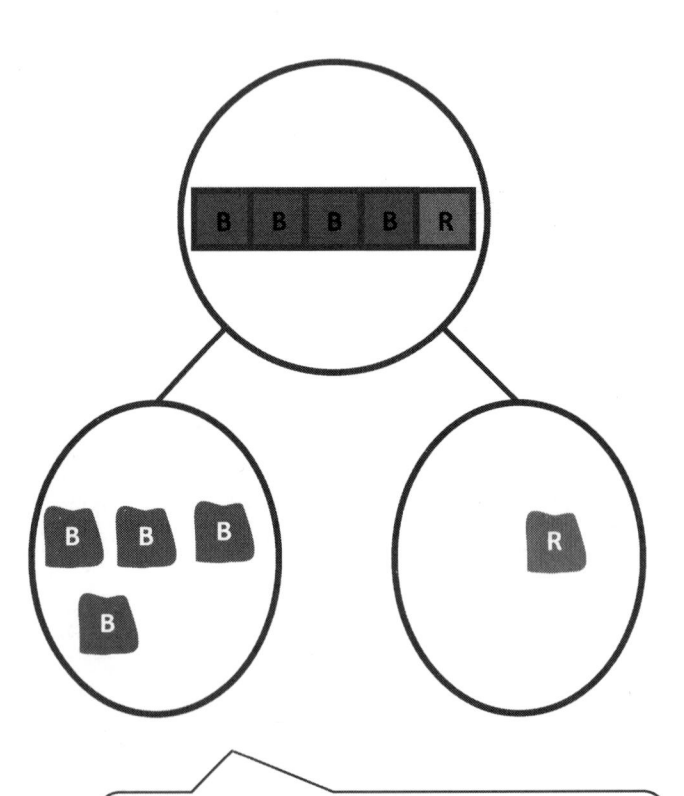

Lesson 2: Model composition and decomposition of numbers to 5 using fingers and linking cubes sticks.

GK-M4-Lesson 3

Fill in the number bond to match the domino.

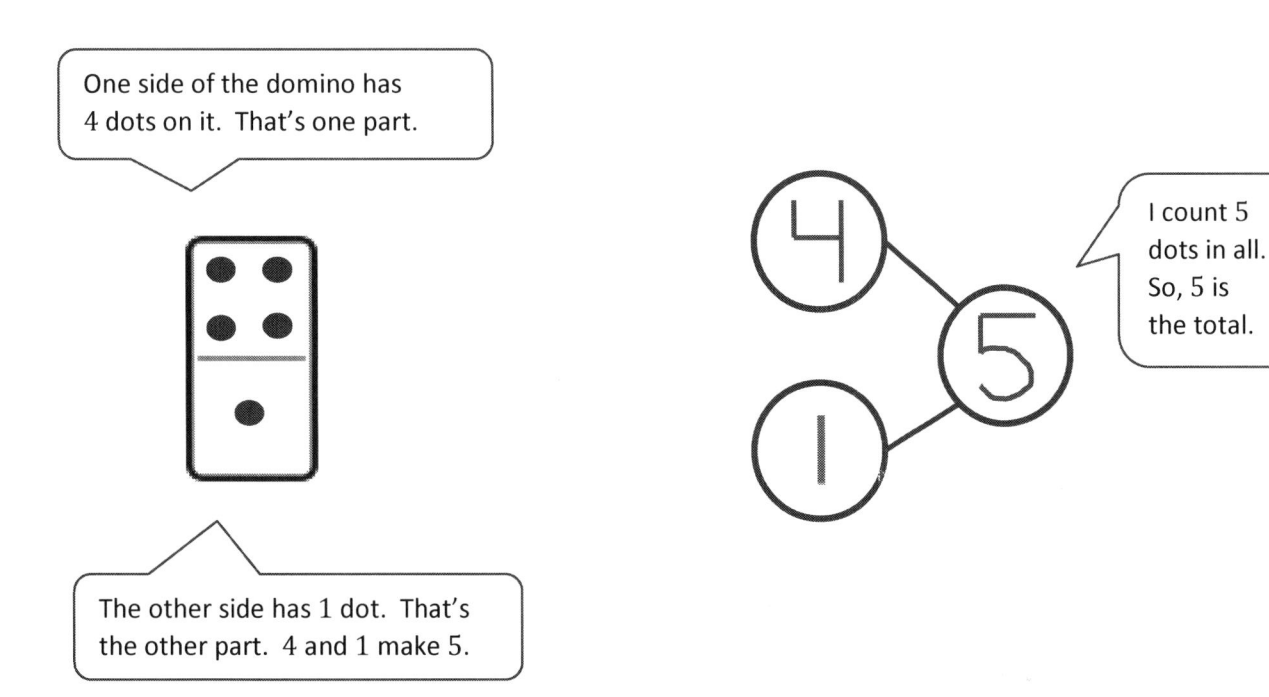

One side of the domino has 4 dots on it. That's one part.

The other side has 1 dot. That's the other part. 4 and 1 make 5.

I count 5 dots in all. So, 5 is the total.

Fill in the domino with dots, and fill in the number bond to match.

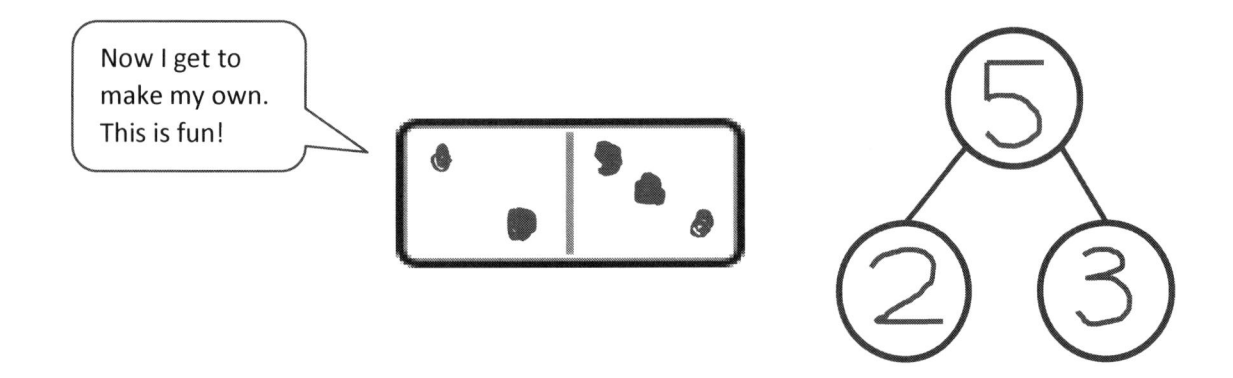

Now I get to make my own. This is fun!

Homework Helper — A Story of Units — K•4

GK-M4-Lesson 4

Finish the number bond. Finish the sentence.

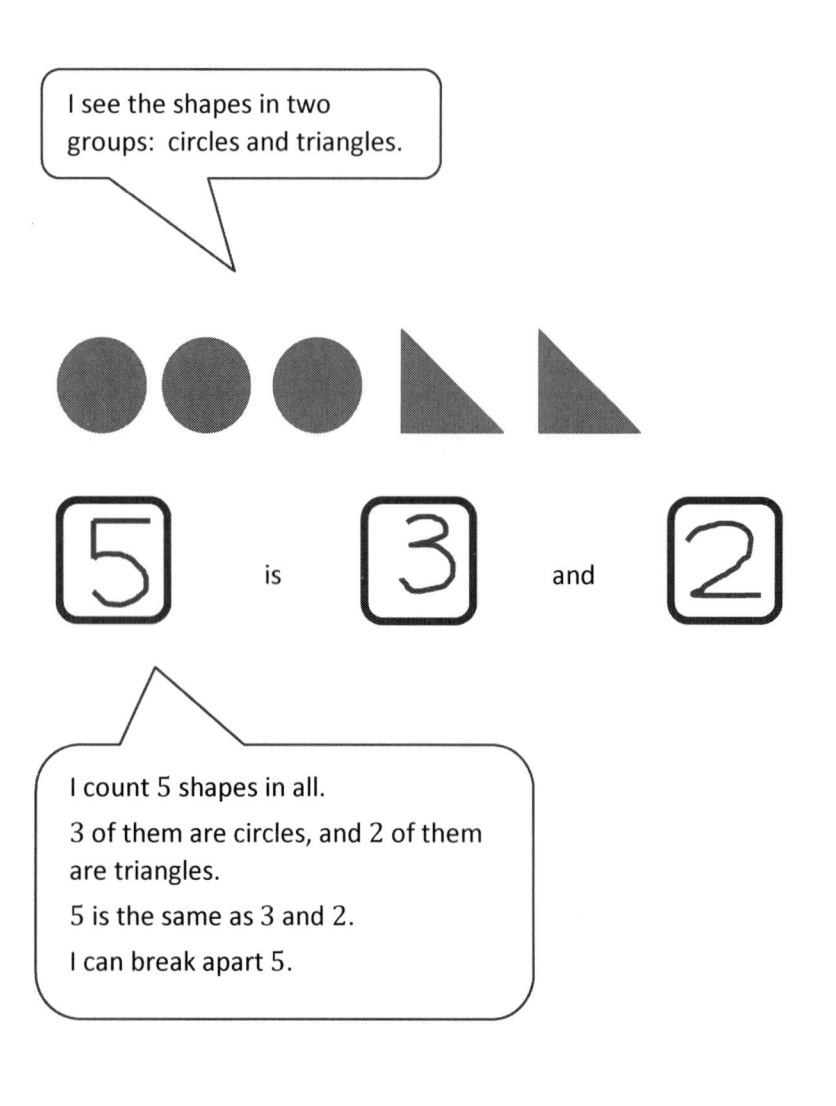

I see the shapes in two groups: circles and triangles.

● ● ● ▲ ▲

[5] is [3] and [2]

(5)
 / \
(3) (2)

I count 5 shapes in all.
3 of them are circles, and 2 of them are triangles.
5 is the same as 3 and 2.
I can break apart 5.

Let me tell you how my number bond matches the picture.
5 is the number of shapes in all.
3 is the number of circles, and 2 is the number of triangles.
I can break apart 5.

Lesson 4: Represent decomposition story situations with drawings using numeric number bonds.

Homework Helper — A Story of Units

GK-M4-Lesson 5

Tell a story about the picture. Fill in the number bond and the sentence to match your story.

> There are 4 happy faces and 1 sad face. There are 5 faces altogether. I can make 5.

4 and 1 make 5

> Let me tell you how my number bond matches the picture.
>
> 4 is the number of happy faces, and 1 is the number of sad faces.
>
> When I put together 4 and 1, they make 5.

Lesson 5: Represent composition and decomposition of numbers to 5 using pictorial and numeric number bonds.

GK-M4-Lesson 6

Tell a story. Complete the number bond. Draw pictures that match your story and number bond.

Draw some animals for your story.

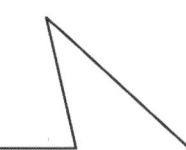

> Listen to my story! At the pet store, I saw 4 animals. 2 of them were cats, and the other 2 were birds.

> 4 is the total. So I draw 4 animals.
> 2 is one of the parts, so I draw 2 cats.
> The other part must be 2 since 4 is the same as 2 and 2.
> I draw 2 birds to make 4 animals in all.

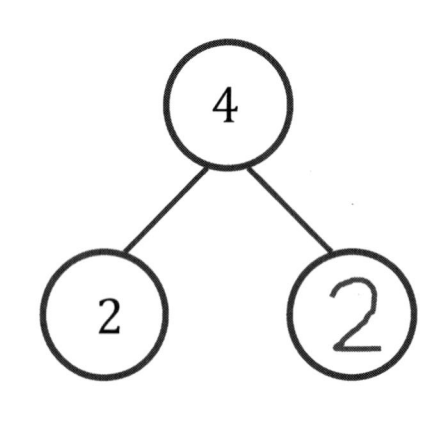

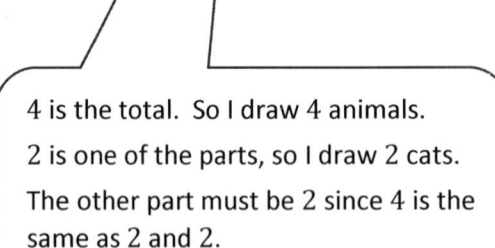

GK-M4-Lesson 7

Look at the shapes. Make 2 different number bonds. Tell an adult about the numbers you put in the number bonds.

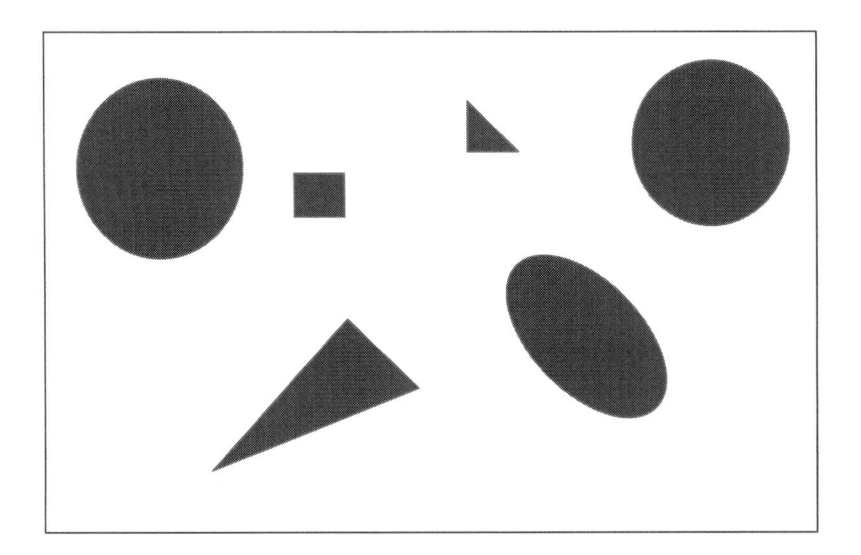

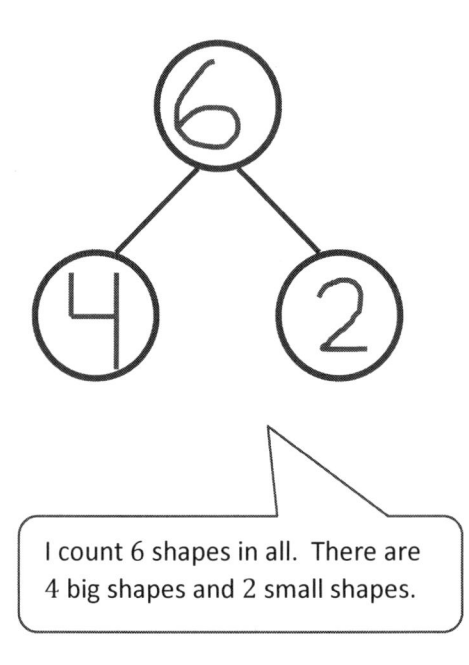

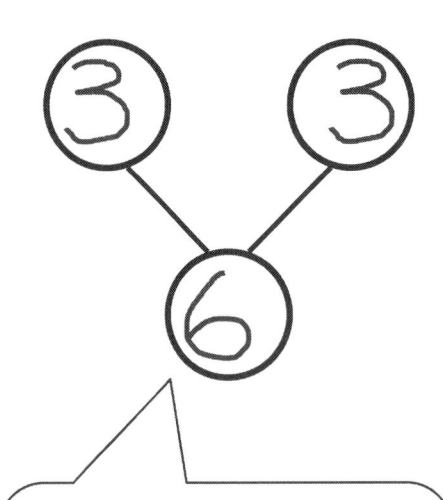

I count 6 shapes in all. There are 4 big shapes and 2 small shapes.

I see shapes with curves and shapes with points. Let me count them. There are 3 curved shapes and 3 pointy shapes.

My number bonds show different ways to break apart 6.

GK-M4-Lesson 8

The squares represent cube sticks. Color some cubes red and the rest blue. Fill in the number bond and sentence to match.

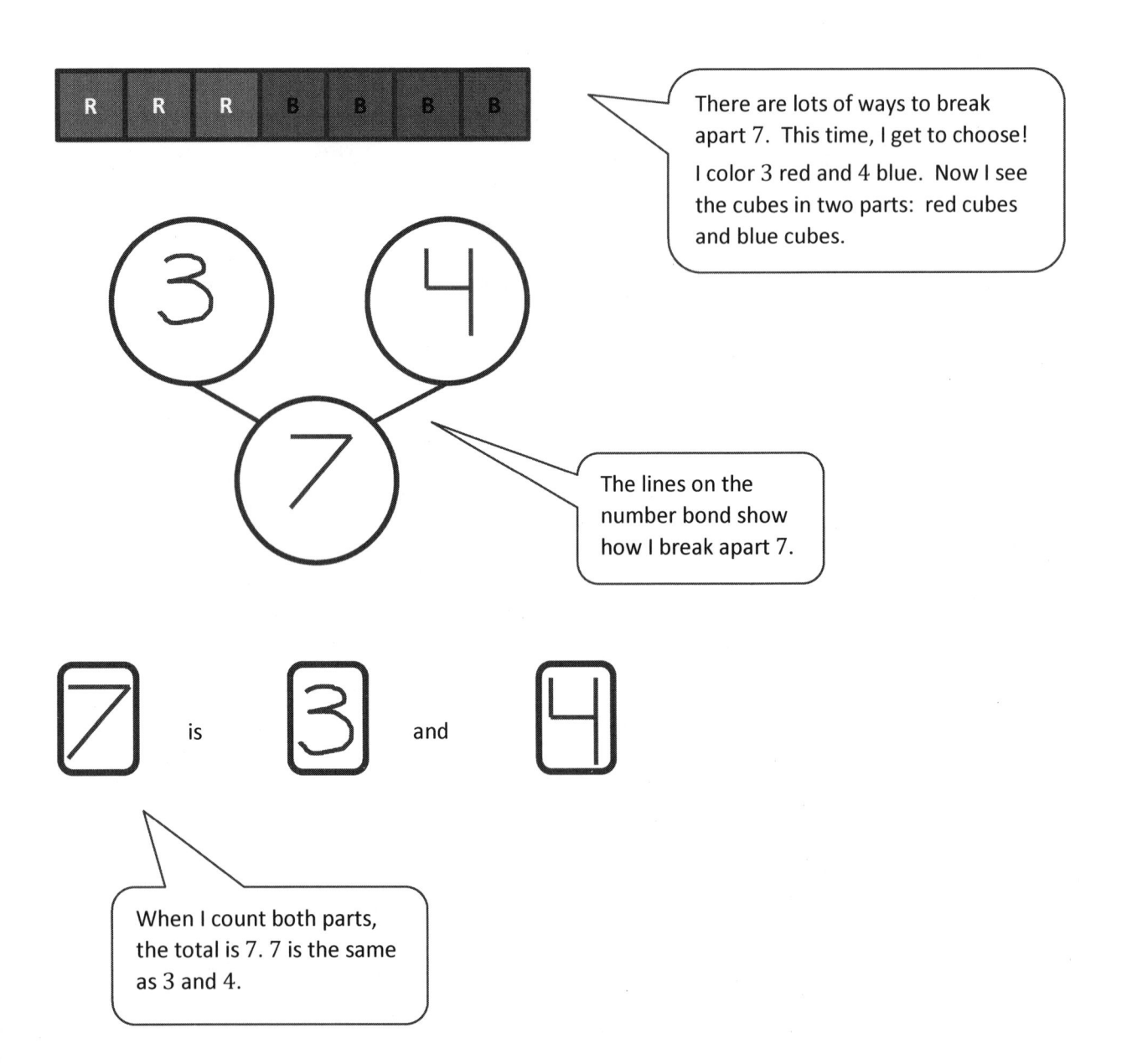

| R | R | R | B | B | B | B |

> There are lots of ways to break apart 7. This time, I get to choose! I color 3 red and 4 blue. Now I see the cubes in two parts: red cubes and blue cubes.

Number bond: 7 breaks into 3 and 4.

> The lines on the number bond show how I break apart 7.

7 is **3** and **4**

> When I count both parts, the total is 7. 7 is the same as 3 and 4.

Lesson 8: Model decompositions of 7 using a story situation, sets, and number bonds.

Homework Helper

A Story of Units

K•4

GK-M4-Lesson 9

Complete the number bond to match the domino.

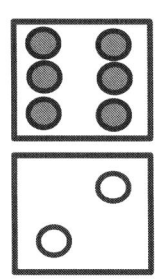

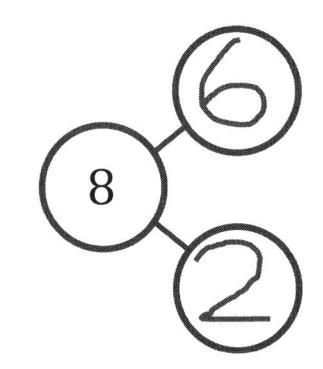

Let me tell you how my number bond and domino match.

8 tells how many dots in all.

6 is the number of grey dots.

2 is the number of white dots.

Draw a line to make 2 groups of dots. Fill in the number bond.

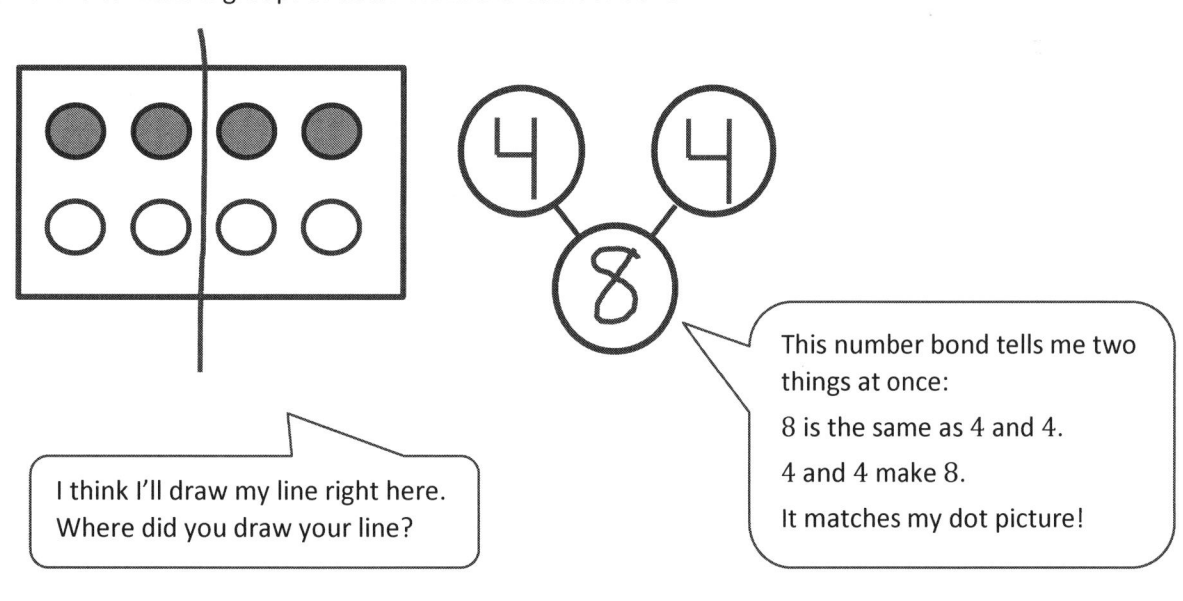

I think I'll draw my line right here. Where did you draw your line?

This number bond tells me two things at once:

8 is the same as 4 and 4.

4 and 4 make 8.

It matches my dot picture!

Lesson 9: Model decompositions of 8 using a story situation, arrays, and number bonds.

GK-M4-Lesson 10

The squares below represent cubes. Color 7 cubes green and 1 blue. Fill in the number bond.

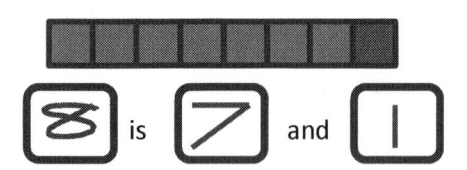

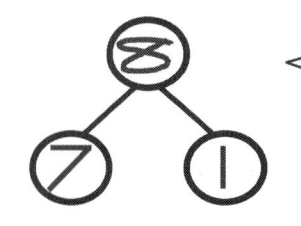

[8] is [7] and [1]

Number bond: 8 → 7 and 1

> The whole stick has 8 cubes. The parts are 7 and 1.

Color 6 cubes green and 2 blue. Fill in the number bond.

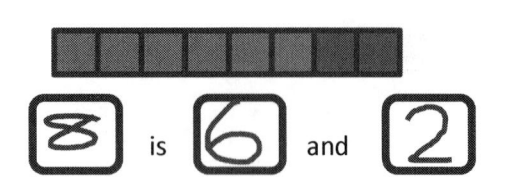

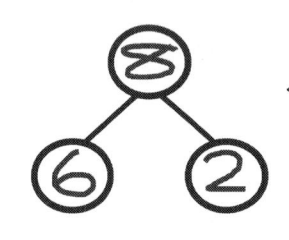

[8] is [6] and [2]

Number bond: 8 → 6 and 2

> This number bond tells 4 things:
> 8 is 6 and 2.
> 8 is 2 and 6.
> 6 and 2 make 8.
> 2 and 6 make 8.

Color some cubes green and the rest blue. Fill in the number bond.

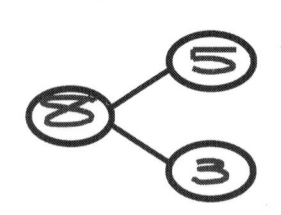

[8] is [5] and [3]

Number bond: 8 → 5 and 3

> Sometimes the whole is on the side.
> The lines show how I took apart 8.

Lesson 10: Model decompositions of 6–8 using linking cube sticks to see patterns.

GK-M4-Lesson 11

These squares represent cubes. Color 5 cubes green and 1 blue. Fill in the number bond.

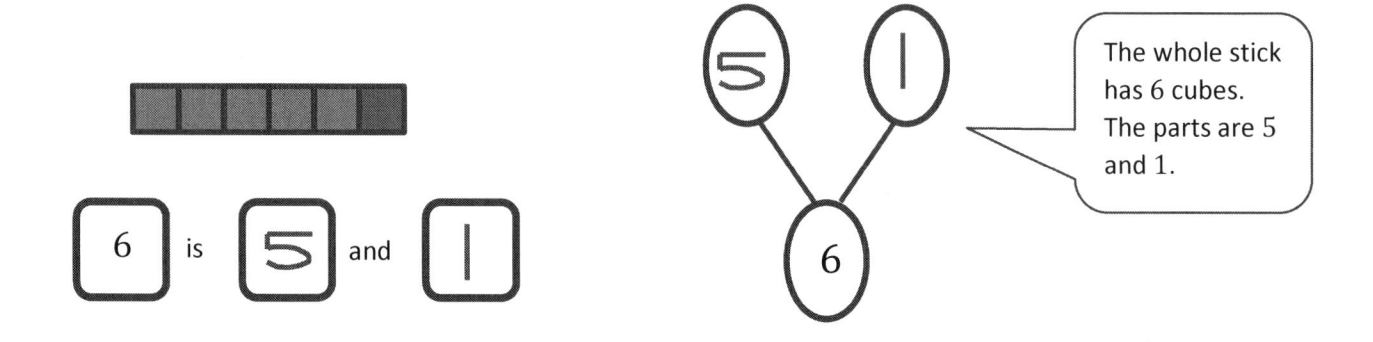

6 is 5 and 1

> The whole stick has 6 cubes. The parts are 5 and 1.

Color 5 cubes green and 2 blue. Fill in the number bond.

7 is 5 and 2

> The whole can be on the top, bottom, or sides. The lines show how the parts go together.

Lesson 11: Represent decompositions for 6–8 using horizontal and vertical number bonds.

GK-M4-Lesson 12

Fill in the number bond to match the squares.

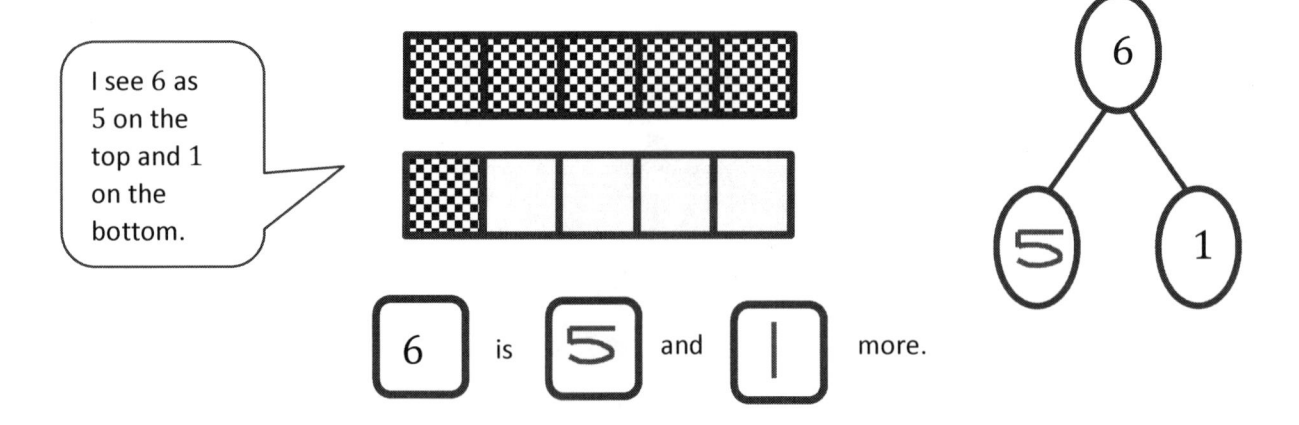

> I see 6 as 5 on the top and 1 on the bottom.

6 is 5 and 1 more.

Number bond: 6 → 5, 1

Color 5 squares blue in the first row.
Color 2 squares red in the second row.

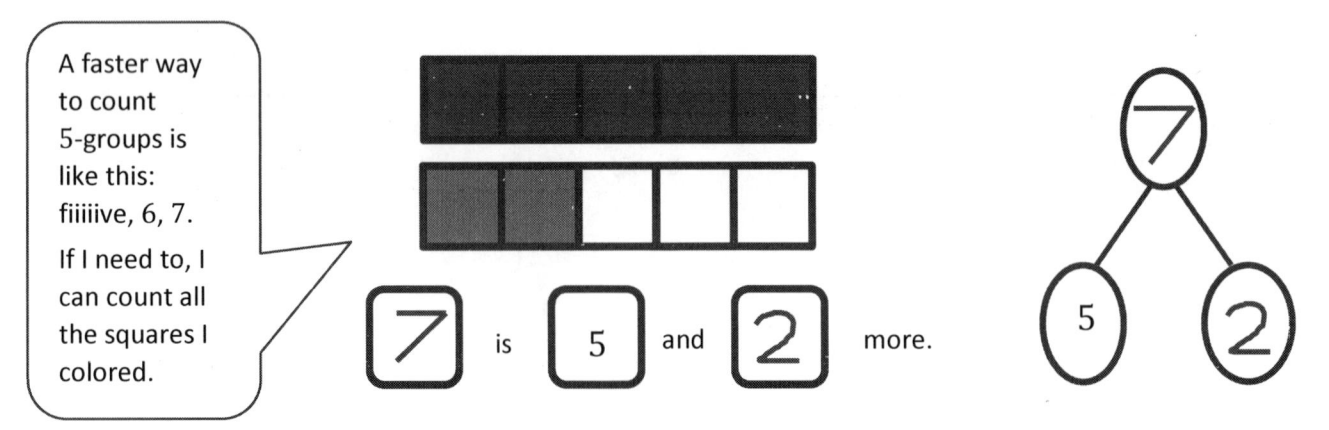

> A faster way to count 5-groups is like this: fiiiiive, 6, 7.
> If I need to, I can count all the squares I colored.

7 is 5 and 2 more.

Number bond: 7 → 5, 2

Lesson 12: Use the 5-groups to represent the 5 + n pattern to 8.

Homework Helper

A Story of Units   K•4

GK-M4-Lesson 13

There are 3 monkeys and 3 elephants. All 6 animals are going into the circus tent. Fill in the number sentence and the number bond.

> This story starts with the parts and ends with the whole.
> I'll write my number sentences that way, too!

3 and 3 is 6

3 + 3 = 6

(Number bond: 3 and 3 make 6)

There are 6 animals. 4 are tigers, and 2 are lions. Fill in the number sentences and the number bond.

> This story is different. It starts with the whole and ends with the parts.
> I'll write my number sentences that way, too!

6 is 4 and 2

6 = 4 + 2

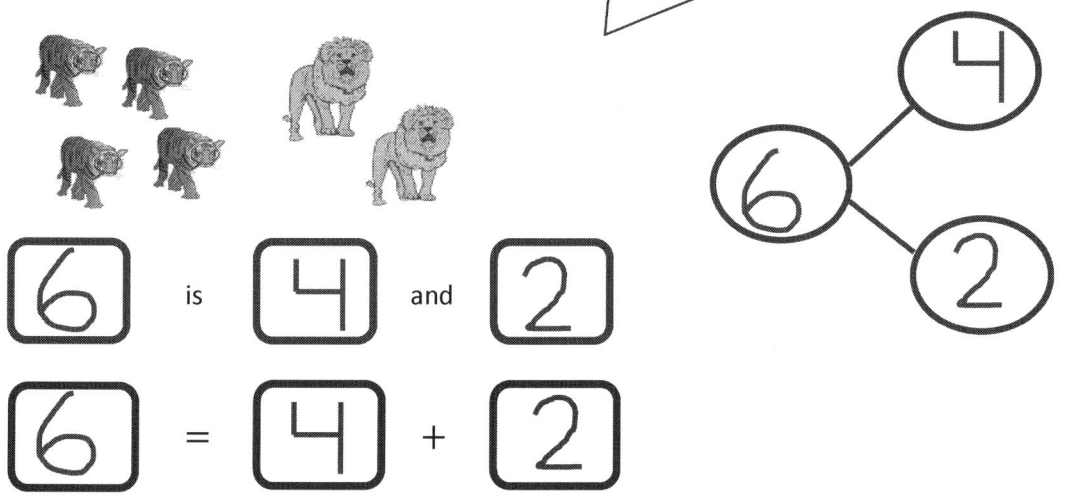

(Number bond: 6 is 4 and 2)

Lesson 13: Represent decomposition and composition addition stories to 6 with drawings and equations with no unknown.

Homework Helper

GK-M4-Lesson 14

There are 7 bears. 3 bears have bowties. 4 bears have hearts. Fill in the number sentences and the number bond.

$7 = 3 + 4$

$3 + 4 = 7$

> I wrote the addition sentences both ways: take apart and put together. My number bond shows that, too!

Lesson 14: Represent decomposition and composition addition stories to 7 with drawings and equations with no unknown.

GK-M4-Lesson 15

There are 8 trees. 5 are palm trees, and 3 are apple trees. Fill in the number sentences and the number bond.

This addition sentence shows that there are 8 trees: 5 of one kind and 3 of another.

$8 = 5 + 3$

This addition sentence shows how the parts go together to make 8.

$5 + 3 = 8$

8 is the whole.
5 and 3 are the parts.

Lesson 15: Represent decomposition and composition addition stories to 8 with drawings and equations with no unknown.

GK-M4-Lesson 16

There are 3 penguins on the ice. 4 more penguins are coming. How many penguins are there?

To find the mystery number, I can count all of the penguins: 1, 2, 3, 4, 5, 6, 7. There are 7 penguins in all!

3 + 4 = [7]

The mystery box is for the number we don't know.
I can trace the mystery box.

Lesson 16: Solve add to with result unknown word problems to 8 with equations. Box the unknown.

GK-M4-Lesson 17

There are 5 hexagons and 2 triangles. How many shapes are there?

> I can add the hexagons and the triangles.
> The total number of shapes is 7.

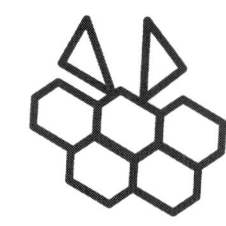

$$\boxed{7} = \boxed{5} + \boxed{2}$$

> I can say this number sentence two ways:
> 7 equals 5 plus 2.
> 7 is the same as 5 and 2.

$$\boxed{5} + \boxed{2} = \boxed{7}$$

> I can say this number sentence two ways:
> 5 plus 2 equals 7.
> 5 and 2 make 7.

Lesson 17: Solve put together with total unknown word problems to 8 using objects and drawings.

GK-M4-Lesson 18

Devin has 6 pencils. He put some in his desk and the rest in his pencil box. Write a number sentence to show how many pencils Devin might have in his desk and pencil box.

> The total is 6. I get to choose how many of each!

$6 = \boxed{5} + \boxed{1}$

> I chose 5 + 1, but I could have written 1 + 5, 4 + 2, 2 + 4, or 3 + 3. There are so many partners to 6.

GK-M4-Lesson 19

Later I'll learn about "minus." For now, I can say that 5 trains take away 1 train is 4 trains.

1 train drove away. Cross out 1. Write how many were left.

4 tells how many are left.

It doesn't matter which one I cross out as long as I cross out 1.

Two Ways to Cross Out

One at a time

All at once

Lesson 19: Use objects and drawings to find how many are left.

GK-M4-Lesson 20

The squares below represent cube sticks. Match the cube stick to the number sentence.

> Let's see. There are 5 squares in the whole stick.
> 2 are crossed out, and 3 are left.
> I can tell about it like this:
> 5 take away 2 is 3.
> Another way is like this:
> 5 minus 2 equals 3.
> We just learned that "minus" is a math word for "take away."

$5 - 3 = 2$

$5 - 1 = 4$

$5 - 2 = 3$

> 5 tells about the whole cube stick.
> Minus 2 tells about the 2 that are crossed off.
> Equals 3 tells about the 3 that are left.
> It's a match!

Homework Helper

GK-M4-Lesson 21

There were 4 oranges. Robin ate 1. Cross out the orange she ate. How many oranges were left? Fill in the boxes.

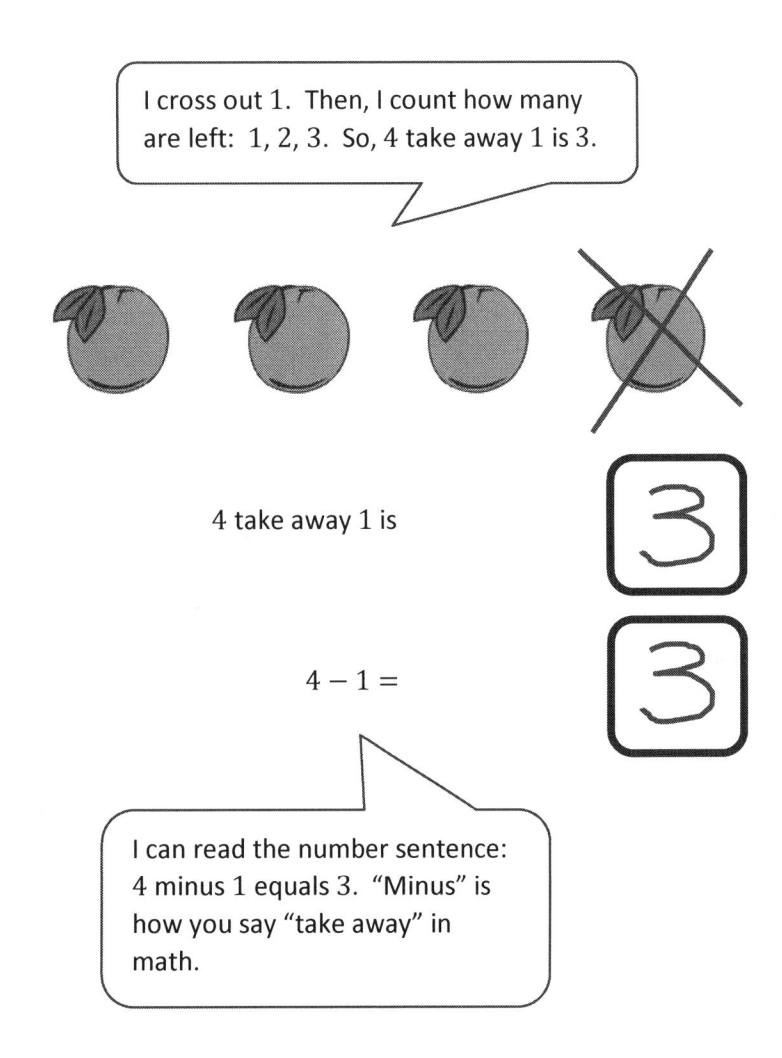

I cross out 1. Then, I count how many are left: 1, 2, 3. So, 4 take away 1 is 3.

4 take away 1 is 3

4 − 1 = 3

I can read the number sentence: 4 minus 1 equals 3. "Minus" is how you say "take away" in math.

Lesson 21: Represent subtraction story problems using objects, drawings, expressions, and equations.

Homework Helper A Story of Units K•4

GK-M4-Lesson 22

Draw 6 hearts. Cross out 2. Fill in the number sentence and the number bond.

♡ ♡ ♡ ♡ ~~♡ ♡~~

> I cross out 2 all at once. That's fast!

> The total, or whole, is 6. That's how many hearts there are in all.
>
> I break apart the group of 6 hearts. Now 2 are crossed out, and 4 are not.
>
> The parts are 2 and 4.

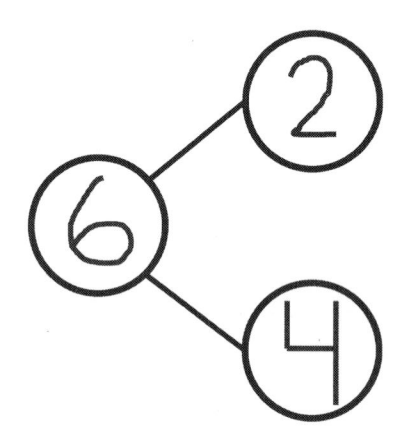

Number bond: 6 → 2 and 4

$$6 - 2 = 4$$

> Crossing out is like taking away, so I subtract. I started with 6 hearts. So my number sentence starts with 6.

Lesson 22: Decompose teen numbers as 10 ones and some ones; compare *some ones* to compare the teen numbers.

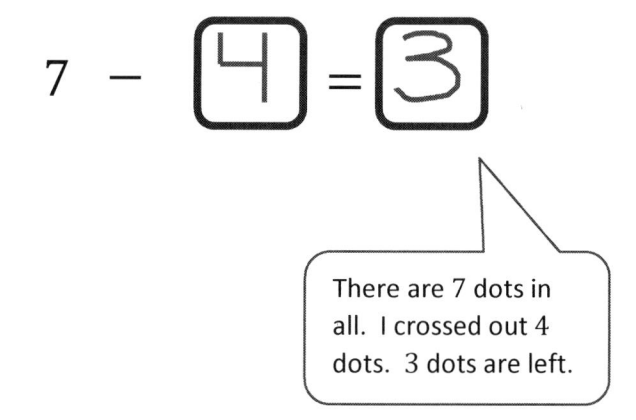

GK-M4-Lesson 23

Draw 7 dots in a 5-group. Cross out 4 dots. Fill in the number sentence and number bond.

I see 4 right here. I cross them out one by one. Which 4 did you cross out?

4 dots are crossed out. 3 dots are not crossed out. 4 and 3 are the parts. 7 is the whole.

Here is another way to draw 7 dots in a 5-group. It's easy to see 7. It's 5 and 2 more!

7 − 4 = 3

There are 7 dots in all. I crossed out 4 dots. 3 dots are left.

Lesson 23: Decompose the number 7 using 5-group drawings by hiding a part, and record each decomposition with a drawing and subtraction equation.

GK-M4-Lesson 24

Here is 8 the 5-group way. Put an X on 3 cubes. How many are left?

Fill in the number sentence and number bond.

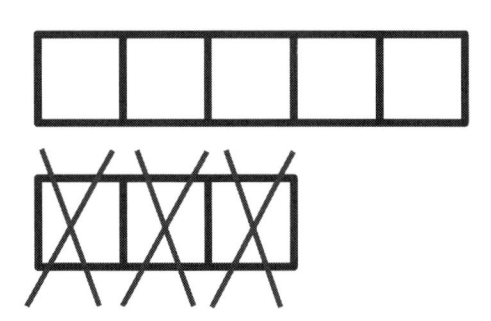

8 − 3 = 5

I did it! My picture, my number bond, and my number sentence all match. I could have crossed out 3 cubes a different way, and it would still match.

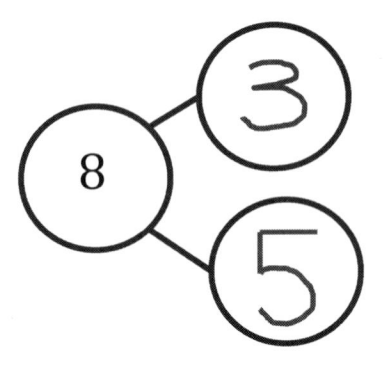

My number bond shows how I broke apart 8. 3 cubes are crossed out. 5 cubes are not crossed out. 8 is the total, or whole. It's like 3 and 5 are hiding inside of 8.

GK-M4-Lesson 25

There are 9 stars. Color some blue and the rest yellow. Fill in the number bond.

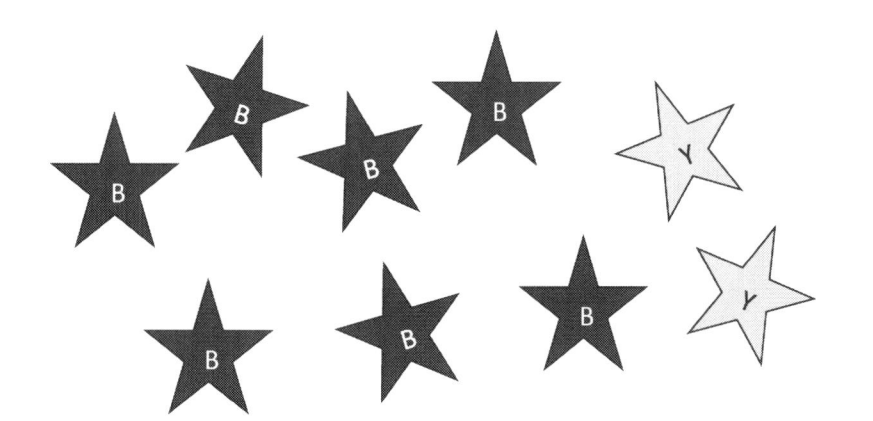

This is fun! I get to choose how many of each color. There are so many ways to break apart 9.

Let me tell you how my number bond goes with my star picture. There are 9 stars in all. That's the total. I color 7 blue and 2 yellow. Those are the parts. When I count all the stars, there are still 9.

Lesson 25: Model decompositions of 9 using a story situation, objects, and number bonds.

Homework Helper — A Story of Units

GK-M4-Lesson 26

The squares below represent cube sticks.
Do the linking cube sticks match the number bond? Circle yes or no.

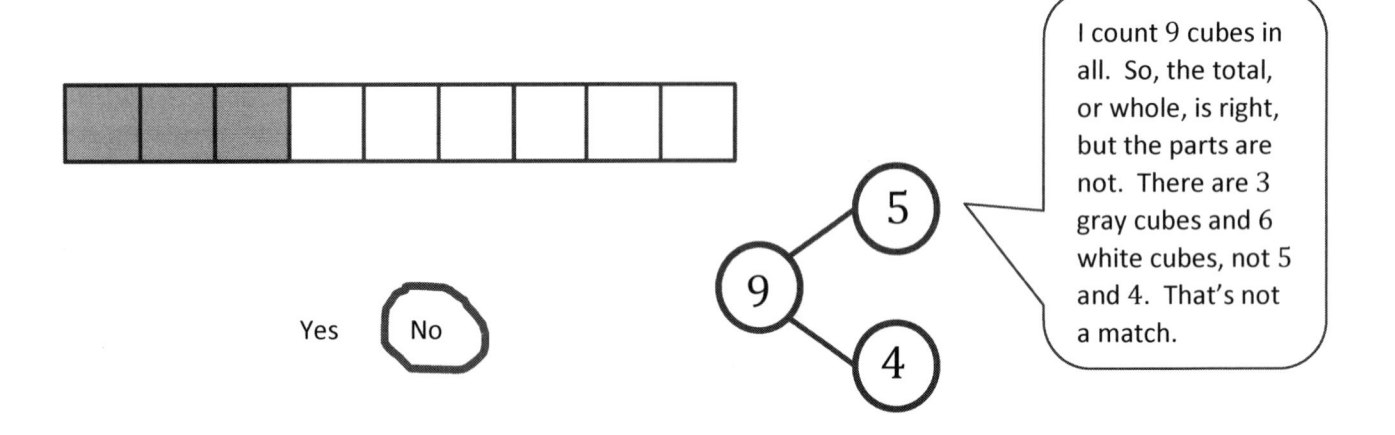

Yes (No)

> I count 9 cubes in all. So, the total, or whole, is right, but the parts are not. There are 3 gray cubes and 6 white cubes, not 5 and 4. That's not a match.

Number bond: 9 → 5 and 4

Make the number bond match the cube stick.

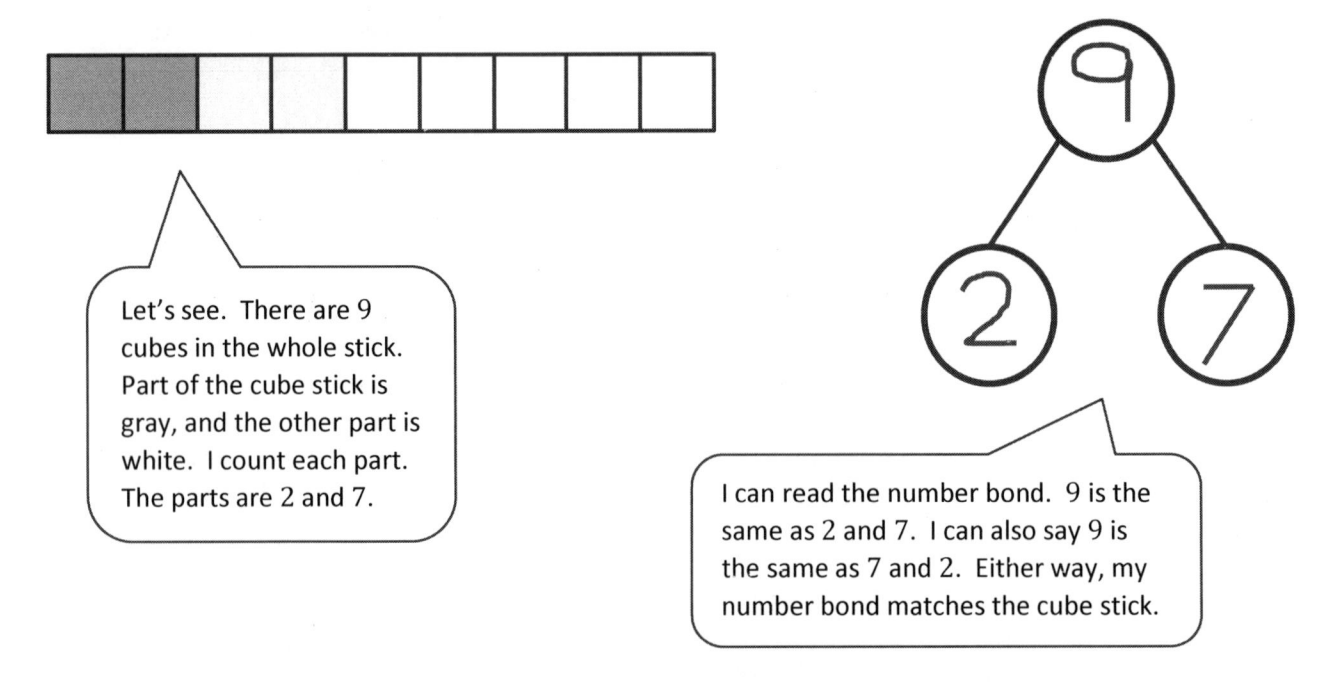

> Let's see. There are 9 cubes in the whole stick. Part of the cube stick is gray, and the other part is white. I count each part. The parts are 2 and 7.

Number bond: 9 → 2 and 7

> I can read the number bond. 9 is the same as 2 and 7. I can also say 9 is the same as 7 and 2. Either way, my number bond matches the cube stick.

Lesson 26: Model decompositions of 9 using fingers, linking cubes, and number bonds.

GK-M4-Lesson 27

Pretend this is your bracelet.

Color some beads red and the rest black. Make a number bond to match.

> Cool! I get to choose how many of each color. I pick 7 red and 3 black. My friend might pick different numbers. No matter what, the total of number of beads on each of our bracelets is still 10.

> The whole, or total, is 10. The parts are 7 and 3. There are 10 beads on the whole entire bracelet. The number 7 is for just the red beads, and the number 3 is for just the black beads.

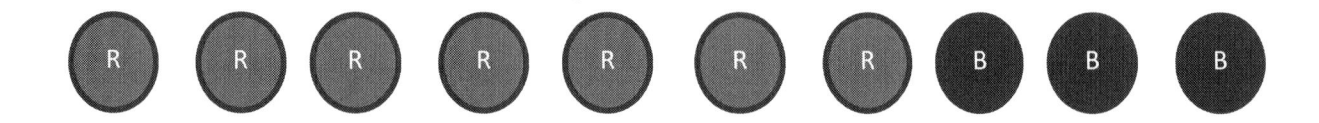

Lesson 27: Model decompositions of 10 using a story situation, objects, and number bonds.

Homework Helper — A Story of Units — K•4

GK-M4-Lesson 28

Write a number bond to match each domino.

It's easy to break apart numbers with dominoes. Just count the number of dots on each side to get the parts.

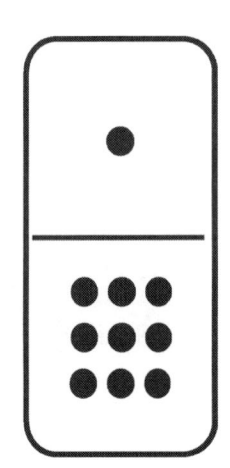

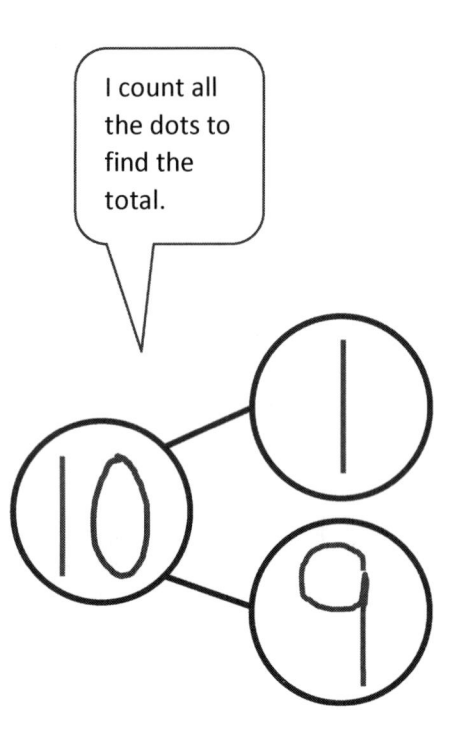

I count all the dots to find the total.

10
1
9

There are so many ways to break apart 10. This one is just like the fingers on both of my hands!

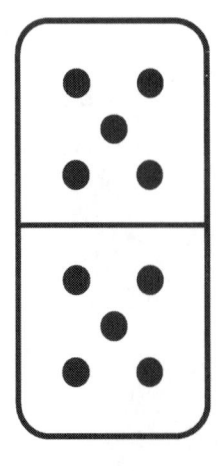

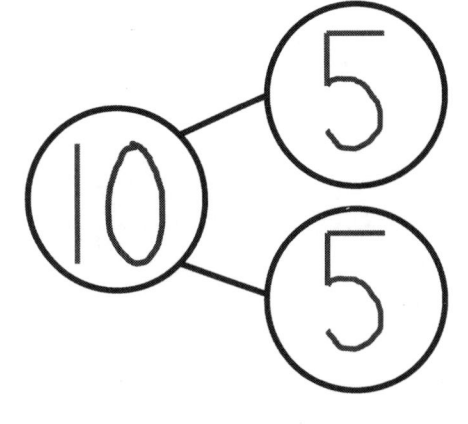

10
5
5

Lesson 28: Model decompositions of 10 using fingers, sets, linking cubes, and number bonds.

Homework Helper — A Story of Units

GK-M4-Lesson 29

Rosey found 8 paintbrushes and 1 gluestick. She found 9 art things. Draw the paintbrushes and the glue stick in the 5-group way. Fill in the number sentence.

● ● ● ● ●
● ● ● ○

> I draw two kinds of dots: circles and filled in circles. That way, I can remember what they go with. The filled in circles are for the paintbrushes. The regular circle is for the glue stick.

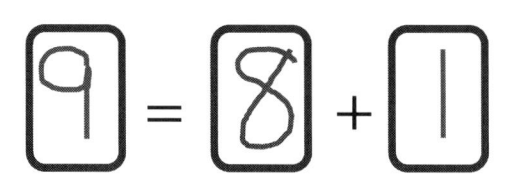

$9 = 8 + 1$

> I can read the number sentence two ways. 9 equals 8 plus 1. Or, 9 is the same as 8 and 1. That helps me understand better!

Jack needs a snack. He found 9 pieces of fruit. 5 were apples, and 4 were oranges. Draw the apples and oranges in the 5-group way.
Fill in the number sentence.

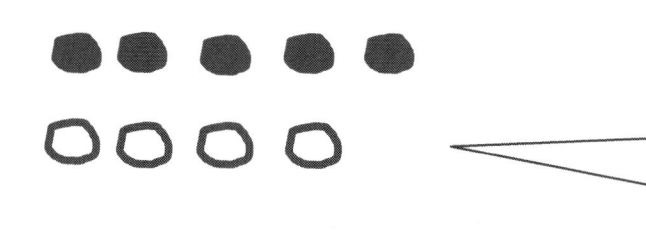

● ● ● ● ●
○ ○ ○ ○

> To draw the 5-group way, I draw dots on the top row, from left to right. 9 is 5 and 4, so I draw 5 dots on the top, and 4 on the bottom.

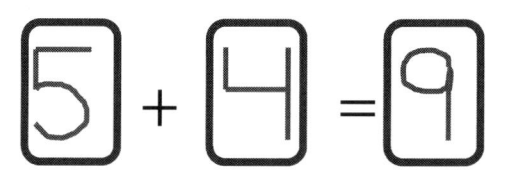

$5 + 4 = 9$

Lesson 29: Represent pictorial decomposition and composition addition stories to 9 with 5-group drawings and equations with no unknown.

Homework Helper — A Story of Units — K•4

GK-M4-Lesson 30

Ming saw 10 animals at the pet store. She saw 6 fish and 4 turtles. Draw the animals in the 5-group way.

> 6 is 5 and 1, so I draw 5 dots on the top and 1 on the bottom.
>
> To draw the other part, 4, I just filled in the rest of the 5-group. That's easy. It makes 10.

6 + 4 = 10

Make 2 groups. Circle 1 of the groups. Write a number sentence to match. Find as many partners of 10 as you can.

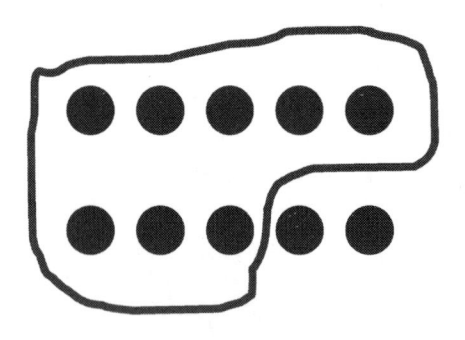

> I can use my imagination to make 2 groups. I pretend the dots are crayons. 8 are in the box, and 2 are on the table.

> Listen to me say the number sentence. 10 equals 8 plus 2. Or, 10 is the same as 8 and 2. Both ways are right!

10 = 8 + 2

Lesson 30: Represent pictorial decomposition and composition addition stories to 10 with 5-group drawings and equations with no unknown.

GK-M4-Lesson 31

Draw the story. Fill in the number sentence.

Ke'Azia has 6 chocolate chip cookies and 3 sugar cookies. How many cookies does she have altogether?

> I can count all of them: 1, 2, 3, 4, 5, 6, 7, 8, 9.
>
> A faster way is siiiiiix, 7, 8, 9. That's how first graders do it!

$\underline{6} + \underline{3} = \underline{9}$

Mario's mother bought juice boxes. 5 were lemonade, and 4 were fruit punch. How many juice boxes did she have in all?

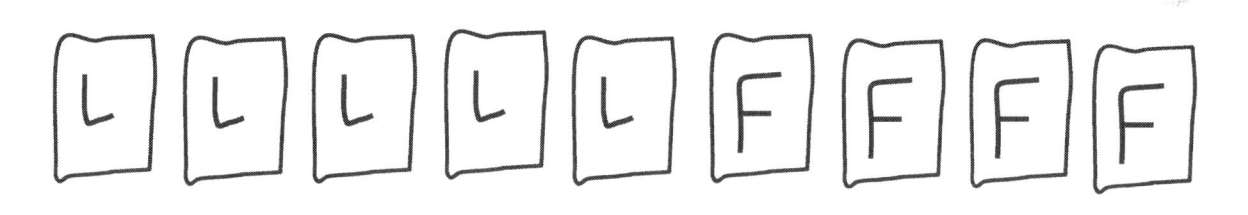

> Math drawings don't have to look like the real thing. I can just put an L, and my teacher will know it's lemonade.

$\underline{5} + \underline{4} = \underline{9}$

Lesson 31: Solve *add to with total unknown* and *put together with total* unknown problems with totals of 9 and 10.

GK-M4-Lesson 32

Anya has 9 stuffed cats. Some are gray, and the rest are white. Show two different ways Anya's cats could look. Fill in the number sentences to match.

I colored this one the 5-group way.

I colored this one a different way.

$9 = \boxed{5} + \boxed{4}$

$9 = \boxed{6} + \boxed{3}$

9 is the same as 5 and 4. It is also the same as 6 and 3. There is more than one way to break apart 9.

Homework Helper — A Story of Units K•4

GK-M4-Lesson 33

Fill in the number sentence to match the story.

There were 10 teddy bears. Cross out 3 bears. There are 7 bears left.

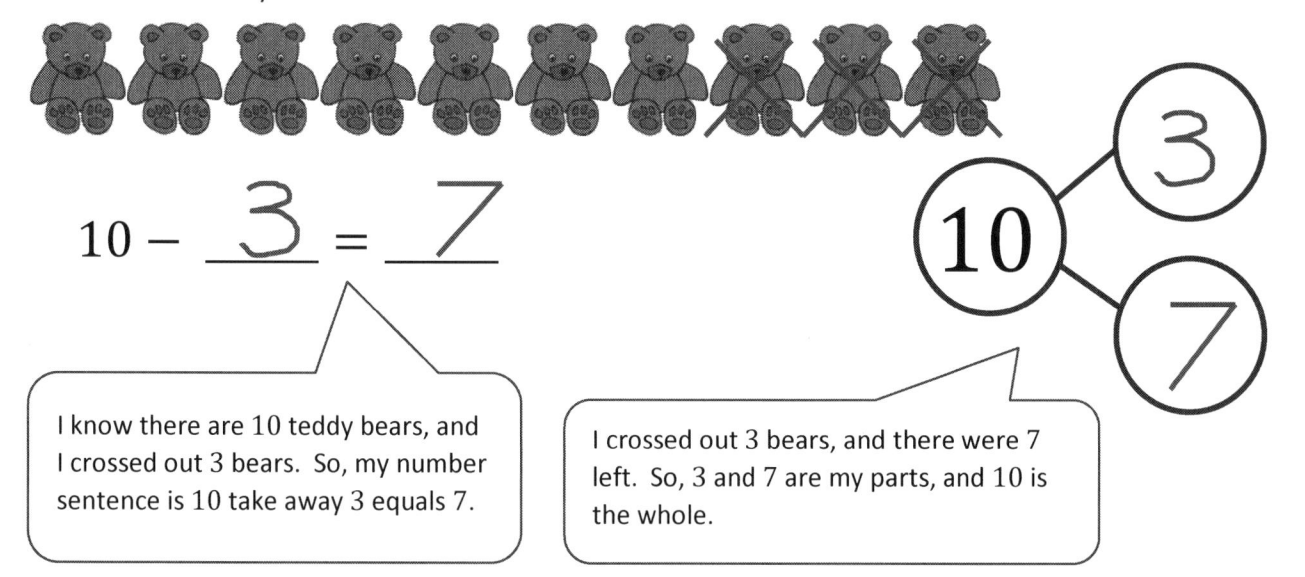

$10 - \underline{3} = \underline{7}$

> I know there are 10 teddy bears, and I crossed out 3 bears. So, my number sentence is 10 take away 3 equals 7.

> I crossed out 3 bears, and there were 7 left. So, 3 and 7 are my parts, and 10 is the whole.

Draw a line from the picture to the number sentence it matches.

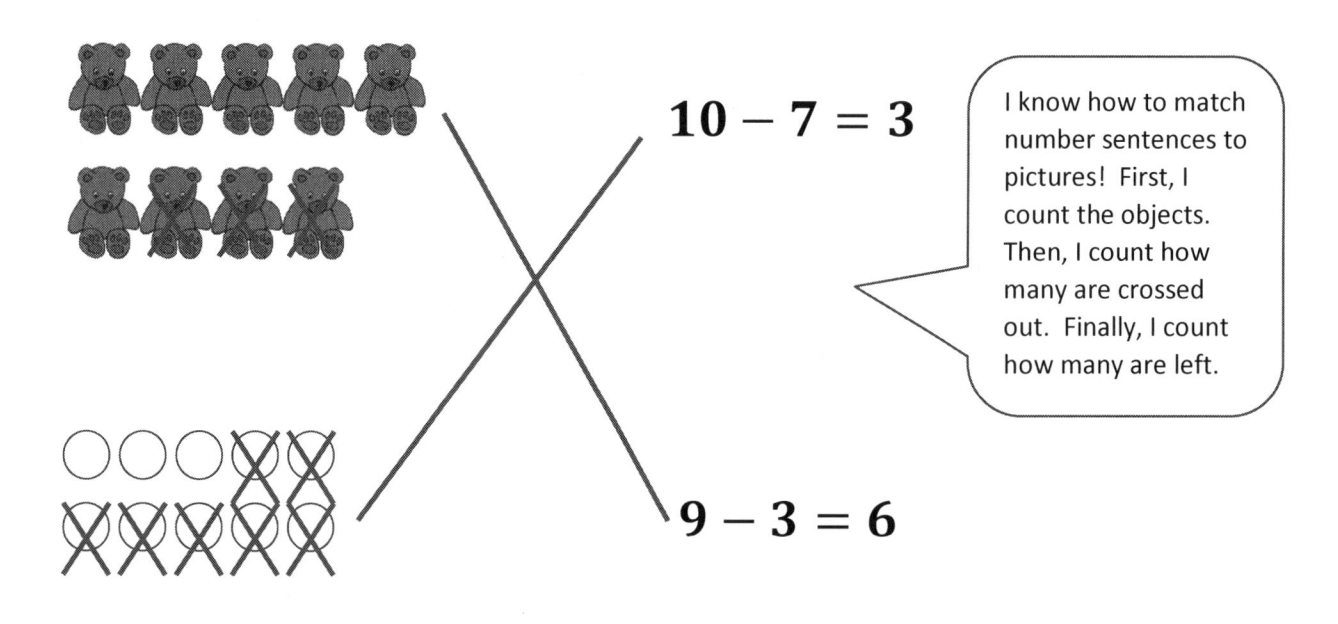

$10 - 7 = 3$

$9 - 3 = 6$

> I know how to match number sentences to pictures! First, I count the objects. Then, I count how many are crossed out. Finally, I count how many are left.

Lesson 33: Solve *take from* equations with no unknown using numbers to 10.

GK-M4-Lesson 34

There were 10 penguins. 4 penguins went back to the ship. Cross out 4 penguins. Fill in the number sentence and the number bond.

10 − 4 = __6__

> There are 10 penguins. I crossed out 4, and there are 6 left. So, 10 take away 4 equals 6.

> I know that 4 penguins are on the ship, and 6 penguins are not on the ship. 4 and 6 are my parts of 10.

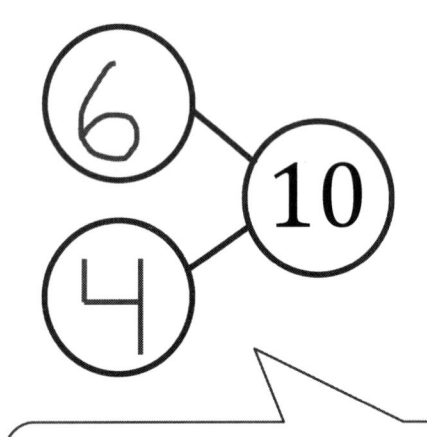

The squares below represent cubes. Count the cubes. Draw a line to break 4 cubes off the train. Fill in the number sentence and the number bond.

> I drew my line to break apart my cube train into parts of 4 and 2. I have 6 cubes. I break off 4 cubes, and I have 2 cubes left!

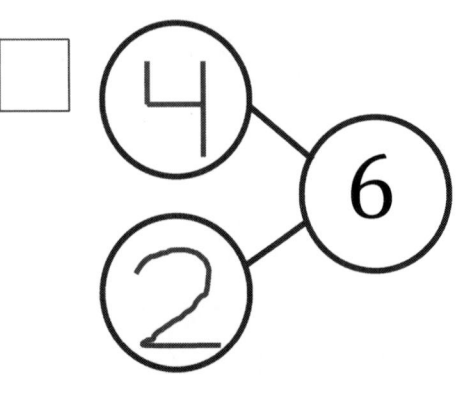

6 − 4 = 2

Lesson 34: Represent subtraction story problems by breaking off, crossing out, and hiding apart.

Homework Helper — A Story of Units

GK-M4-Lesson 35

Cross off the part that goes away. Fill in the number bond and number sentence.

Mary had 9 library books. She returned 2 books to the library. How many books are left?

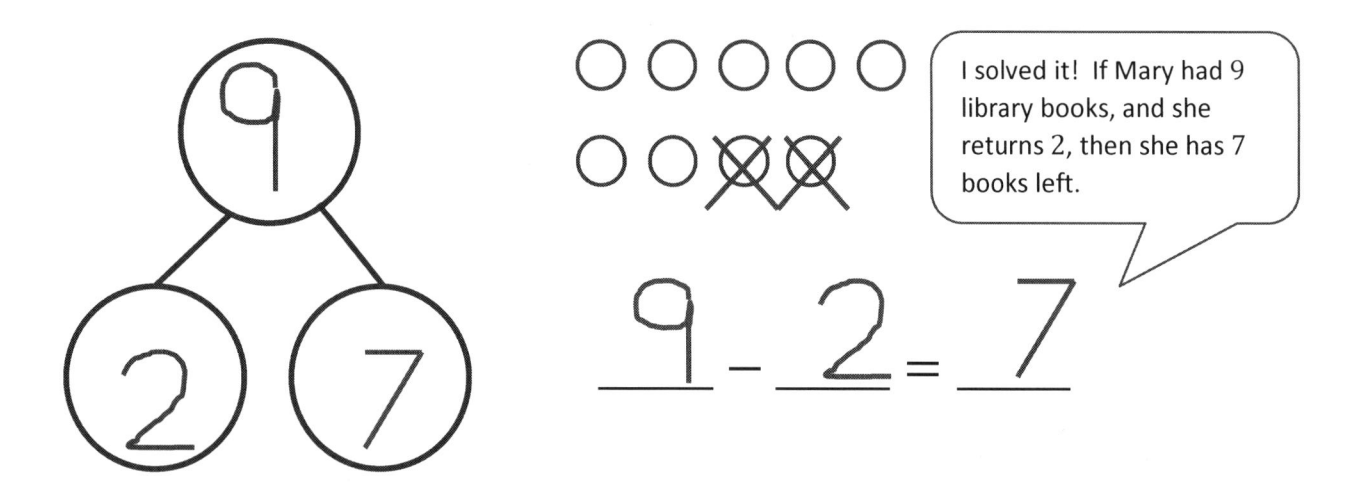

> I solved it! If Mary had 9 library books, and she returns 2, then she has 7 books left.

$9 - 2 = 7$

Make a 5-group drawing to show the story. Cross off the part that goes away. Fill in the number bond and number sentence.

Ryder had 9 pencils. 4 of them broke. How many pencils are left?

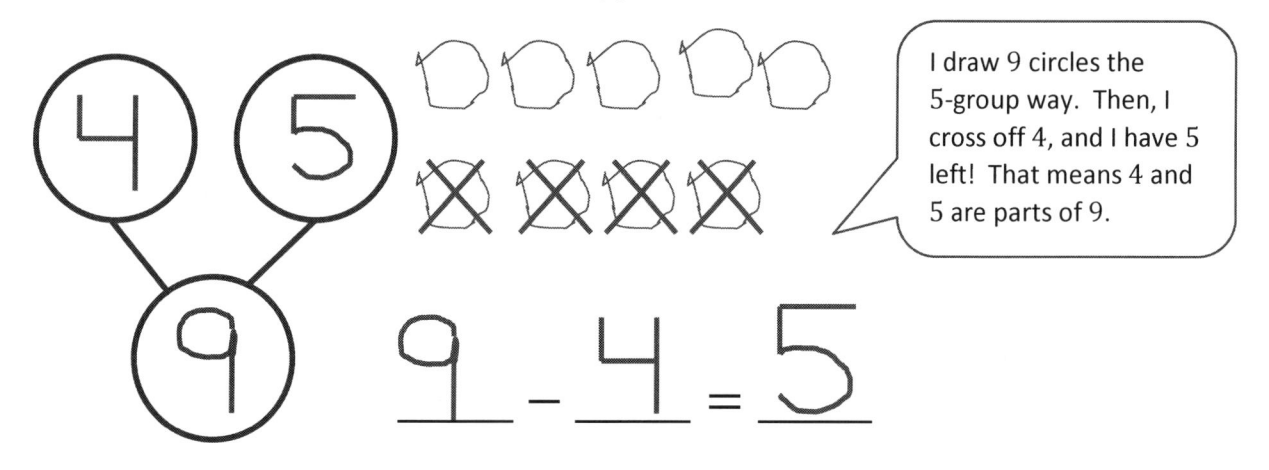

> I draw 9 circles the 5-group way. Then, I cross off 4, and I have 5 left! That means 4 and 5 are parts of 9.

$9 - 4 = 5$

Lesson 35: Decompose the number 9 using 5-group drawings, and record each decomposition with a subtraction equation.

Homework Helper

GK-M4-Lesson 36

Fill in the number bond and number sentence. Cross off the part that goes away.

MacKenzie had 10 buttons on her jacket. 4 buttons broke off her jacket. How many buttons are left on her jacket?

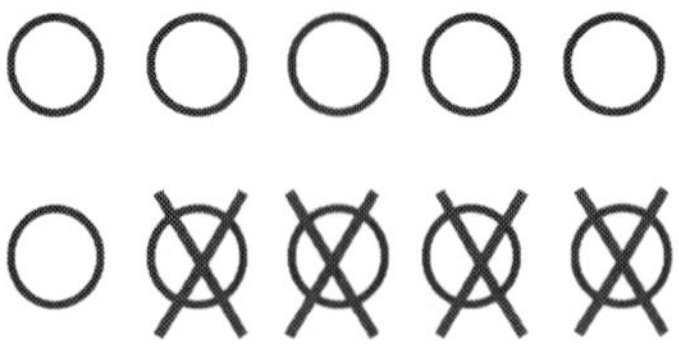

> I know there were 10 buttons on the jacket. 4 broke and fell off. There are 6 buttons left on the jacket. I already knew that 4 and 6 make 10. So, 10 take away 4 is 6.

Number bond: 10 → 6 and 4

$10 - 4 = 6$

Make a 5-group drawing to show the story. Fill in the number bond and number sentence. Cross off the part that goes away.

Bob had 10 toy cars. 3 cars drove away. How many cars are left?

> I made a 5-group drawing to show the cars. 3 drove away, so I crossed out 3. There are 7 cars left.

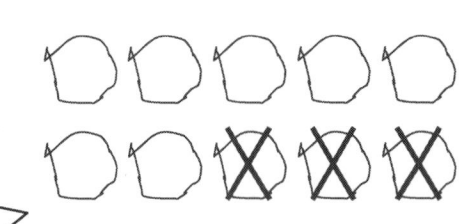

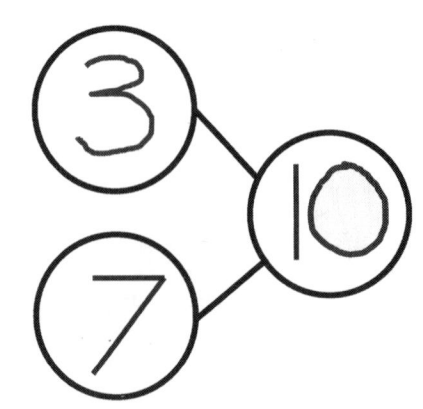

Number bond: 10 → 7 and 3

$10 - 3 = 7$

Lesson 36: Decompose the number 10 using 5-group drawings, and record each decomposition with a subtraction equation.

Homework Helper — A Story of Units

GK-M4-Lesson 37

Listen to each story. Show the story with your fingers on the number path. Then, fill in the number sentence and number bond.

| 1 | 2 | 3 | 4 | 5 | 6 | 7 | 8 | 9 | 10 |

Joey had 7 pennies. He found 2 pennies in the couch. How many pennies does Joey have now?

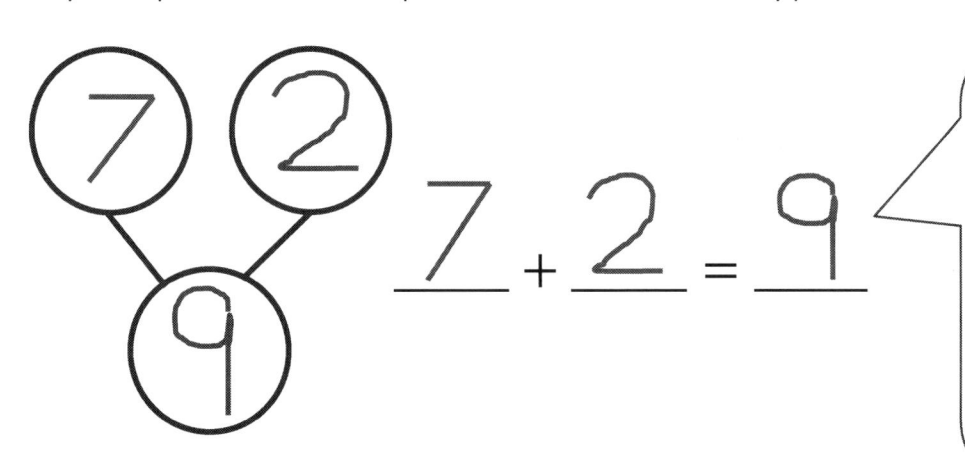

Number bond: 7 and 2 make 9.

$7 + 2 = 9$

> I use the number path to help me solve the problem! I put my finger on 7 because Joey had 7 pennies. He found 2 pennies, so I hop forward 2 on the number path. My fingers stop on the 9. Joey has 9 pennies!

Joey gave the 2 pennies to his dad. How many pennies does Joey have now?

Number bond: 9 splits into 2 and 7.

$9 - 2 = 7$

> I know that Joey has 9 pennies. He gave his dad 2 pennies, so I hop 2 backward on the number path. My fingers stop on the 7. Now, Joey has 7 pennies!

Lesson 37: Add or subtract 0 to get the same number and relate to word problems wherein the same quantity that joins a set, separates.

Homework Helper — A Story of Units — K•4

| 1 | 2 | 3 | 4 | 5 | 6 | 7 | 8 | 9 | 10 |

There were 9 children waiting for the school bus. No more children came to the bus stop. How many children are waiting now?

$\underline{9} + \underline{0} = \underline{9}$

(number bond: 9 with parts 9 and 0)

> I know that 9 children are at the bus stop. I put my finger on the 9 on the number path. No more children came, so my finger doesn't move. There are 9 children waiting at the bus stop.

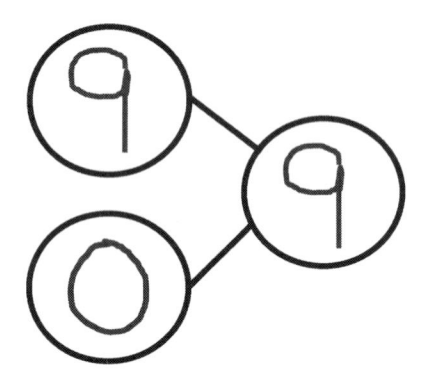

Lesson 37: Add or subtract 0 to get the same number and relate to word problems wherein the same quantity that joins a set, separates.

GK-M4-Lesson 38

Follow the instructions to color the 5-group. Then, fill in the number sentence and number bond to match.

Color 6 squares green and 1 square blue.

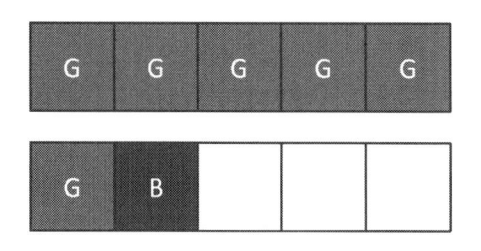

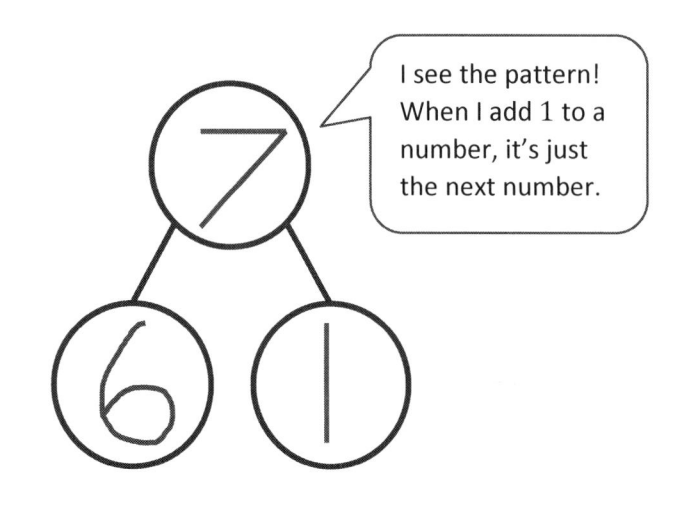

> I see the pattern! When I add 1 to a number, it's just the next number.

$\underline{6} + \underline{1} = \underline{7}$

Color 3 squares green and 1 square blue.

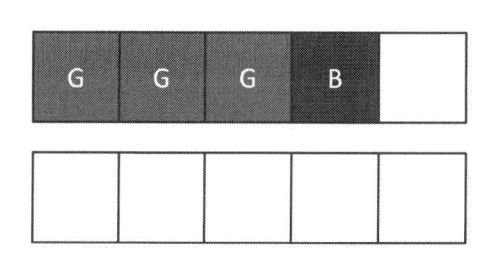

> Adding 1 is easy! 3. 1 more is 4.

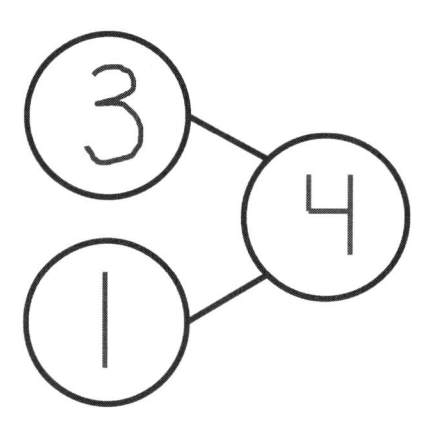

$\underline{3} + \underline{1} = \underline{4}$

Lesson 38: Add 1 to numbers 1–9 to see the pattern of the next number using 5-group drawings and equations.

Homework Helper — A Story of Units

GK-M4-Lesson 39

Draw dots to make 10. Finish the number bonds. Draw a line from the 5-group to the matching number bond.

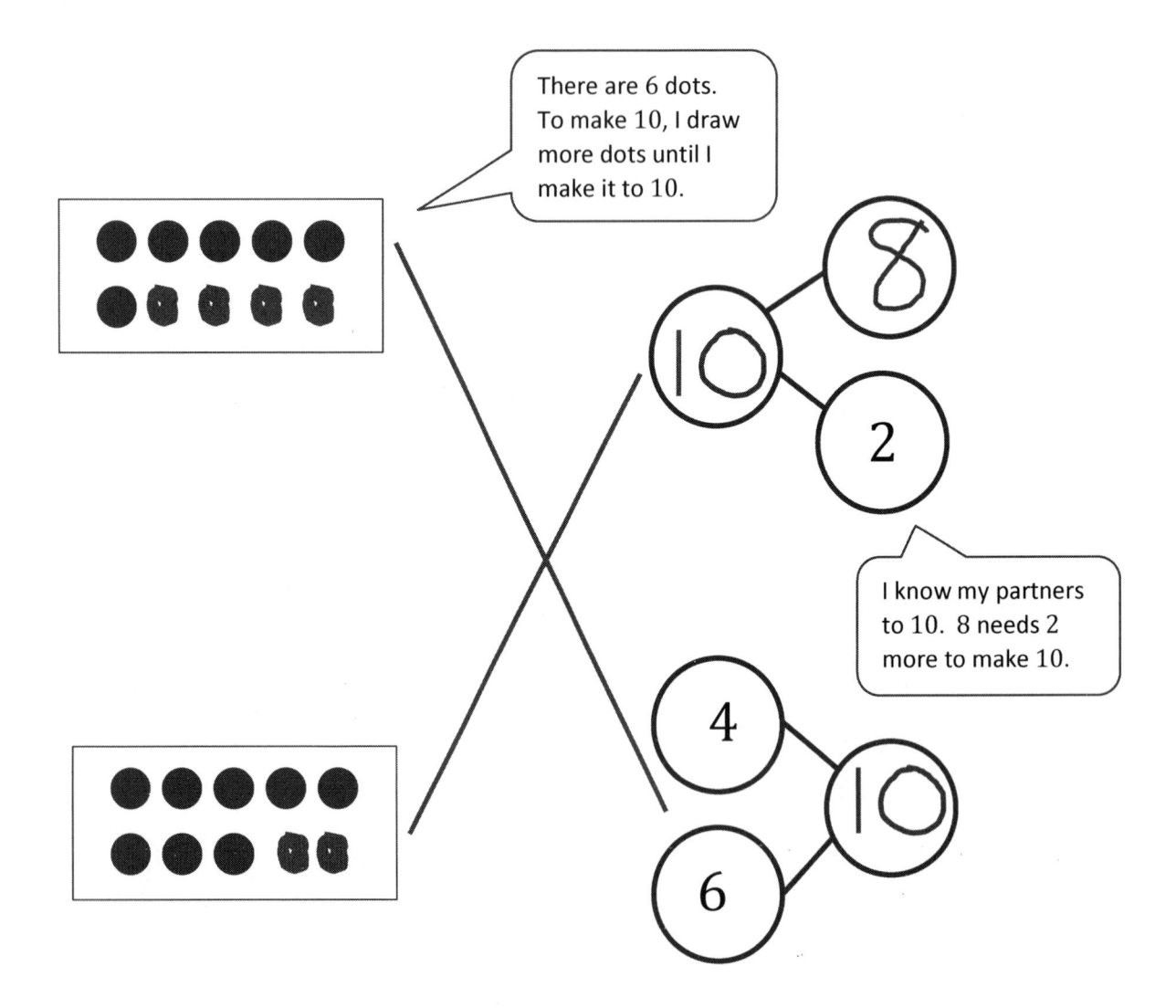

> There are 6 dots. To make 10, I draw more dots until I make it to 10.

> I know my partners to 10. 8 needs 2 more to make 10.

Number bonds: 10 → 8 and 2; 4 and 6 → 10.

Lesson 39: Find the number that makes 10 for numbers 1–9, and record each with a 5-group drawing.

GK-M4-Lesson 40

Color 7 boxes red the 5-group way. Color the rest blue to make 10. Fill in the number sentence.

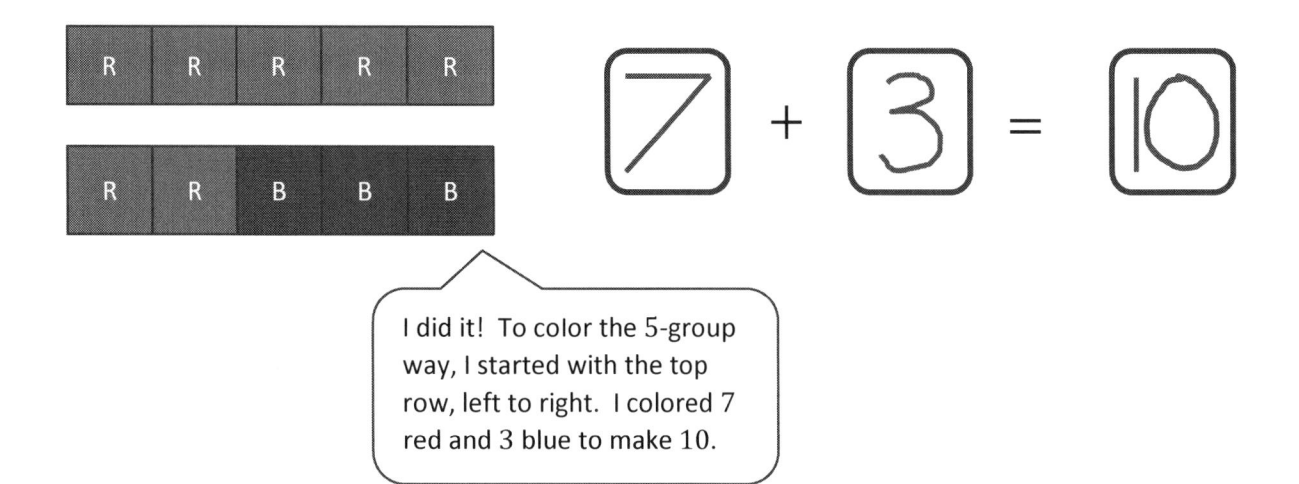

$7 + 3 = 10$

> I did it! To color the 5-group way, I started with the top row, left to right. I colored 7 red and 3 blue to make 10.

Match.

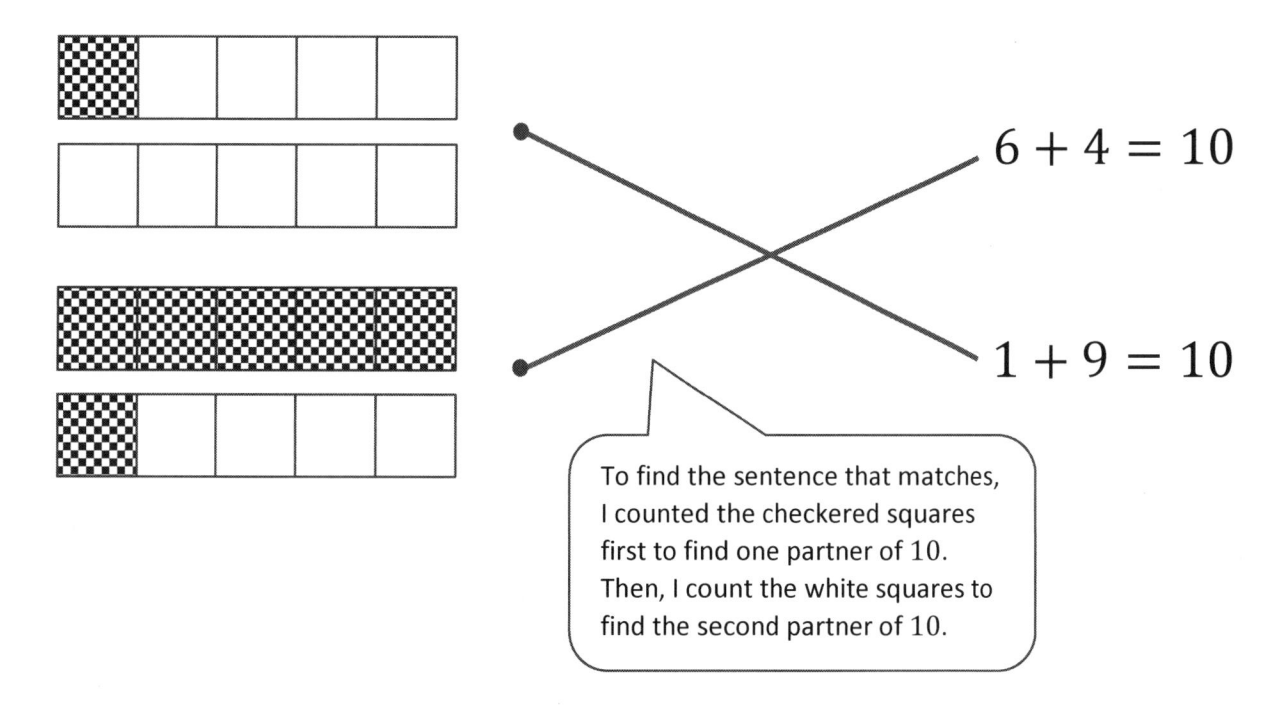

$6 + 4 = 10$

$1 + 9 = 10$

> To find the sentence that matches, I counted the checkered squares first to find one partner of 10. Then, I count the white squares to find the second partner of 10.

Lesson 40: Find the number that makes 10 for numbers 1–9, and record each with an addition equation.

Homework Helper

A Story of Units — K•4

GK-M4-Lesson 41

Complete a number bond and a number sentence for the problem:

Color some blocks orange and the rest yellow to make 10. All of the yellow blocks fell off the table. How many blocks are left?

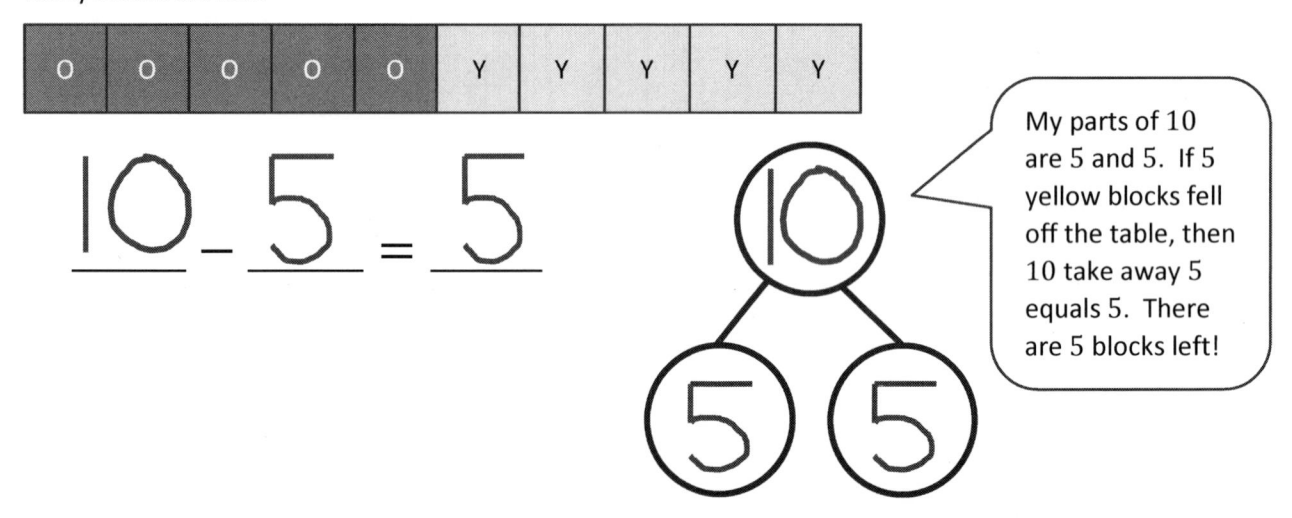

10 − 5 = 5

> My parts of 10 are 5 and 5. If 5 yellow blocks fell off the table, then 10 take away 5 equals 5. There are 5 blocks left!

There were 10 horses in the yard. Some were brown, and some were white. Draw the horses the 5-group way. The brown ones went back into the barn. How many horses were still in the yard? Draw a number bond, and write a subtraction sentence.

10 − 7 = 3

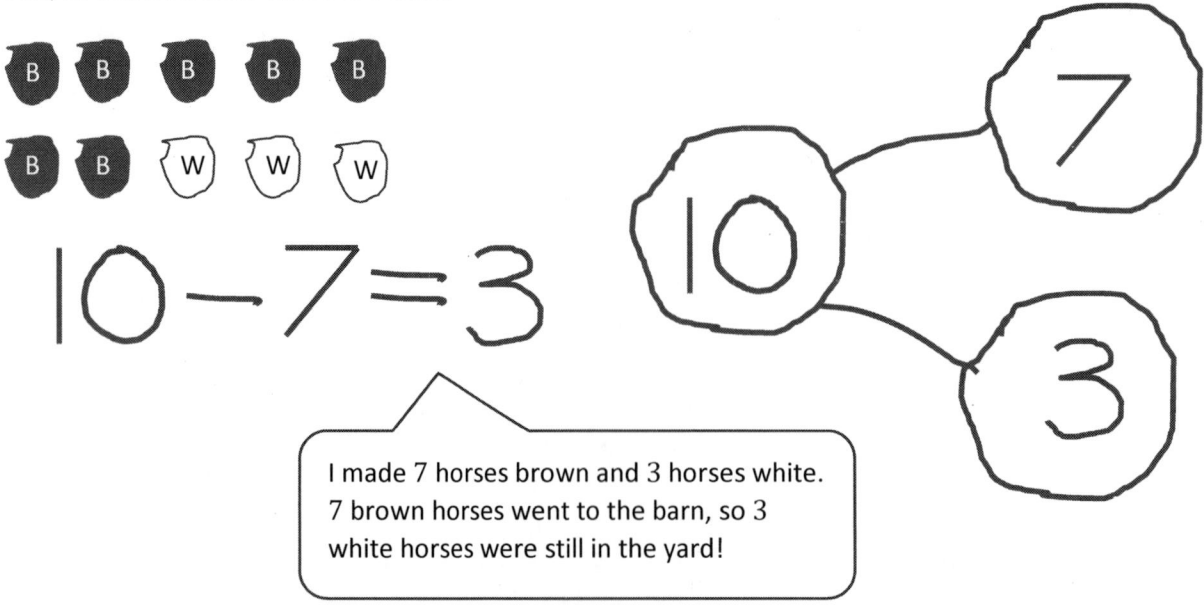

> I made 7 horses brown and 3 horses white. 7 brown horses went to the barn, so 3 white horses were still in the yard!

Lesson 41: Culminating task-choose tools strategically to model and represent a stick of 10 cubes broken into two parts.

Homework Helpers

Grade K
Module 5

GK-M5-Lesson 1

Circle 10. Count the number of times you circled 10 ones. Tell a friend or an adult how many times you circled 10 ones.

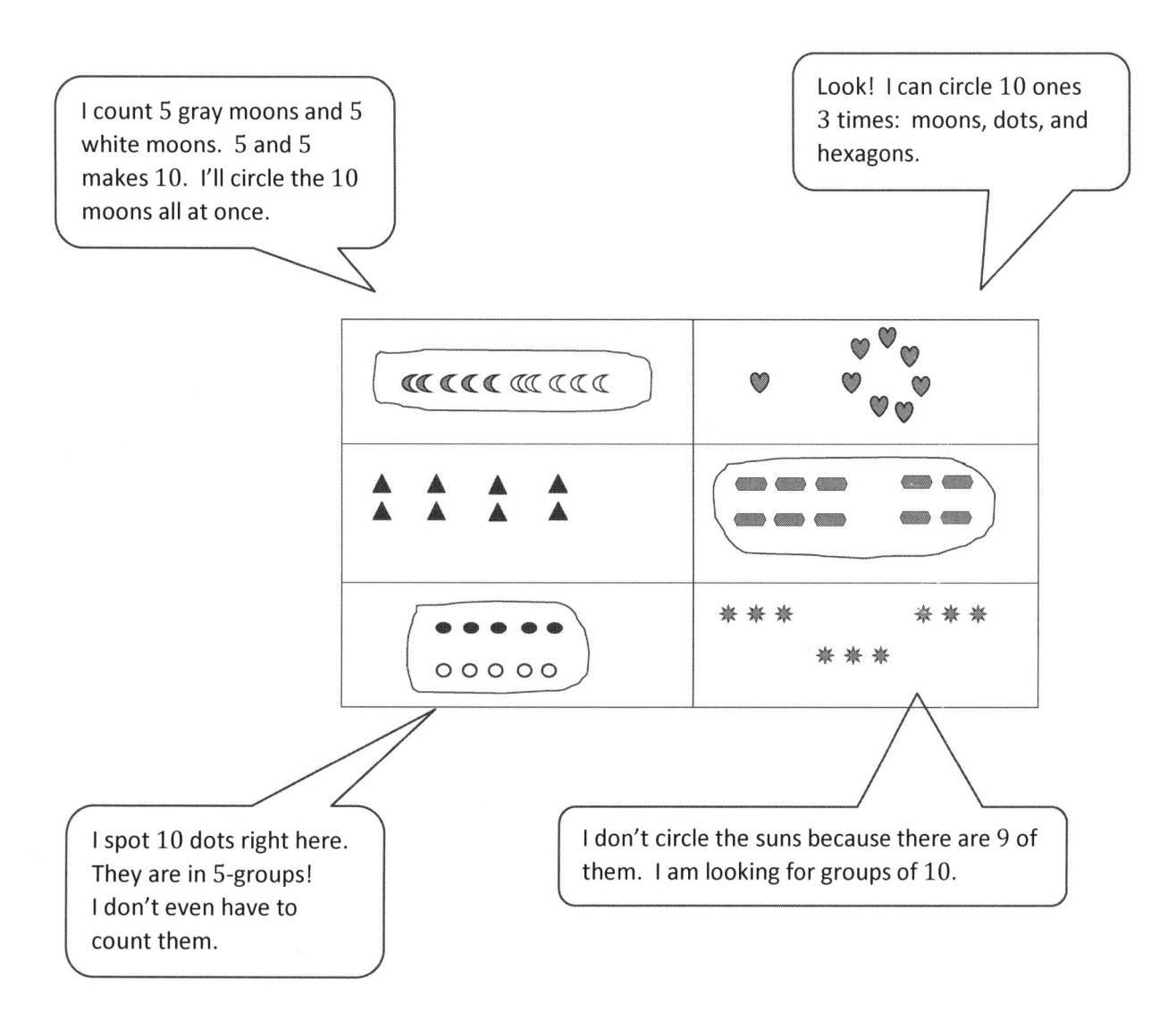

> I count 5 gray moons and 5 white moons. 5 and 5 makes 10. I'll circle the 10 moons all at once.

> Look! I can circle 10 ones 3 times: moons, dots, and hexagons.

> I spot 10 dots right here. They are in 5-groups! I don't even have to count them.

> I don't circle the suns because there are 9 of them. I am looking for groups of 10.

Lesson 1: Count straws into piles of ten; count the piles as 10 ones.

Homework Helper — A Story of Units

GK-M5-Lesson 2

Draw more to show the number.

10 ones and 3 ones

> It's easy to see 10 dots right here. They are in 5-groups! So I just draw 3 more.

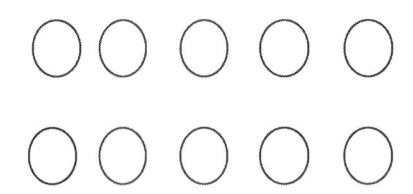

---

10 ones and 6 ones

> There are 9 happy faces already. So I draw 1 more to make 10.

> I draw 6 more off to the side. That makes it easy to see the 10 ones and the 6 ones.

Lesson 2: Count 10 objects within counts of 10 to 20 objects, and describe as 10 ones and __ ones.

Homework Helper — A Story of Units

GK-M5-Lesson 3

Circle 10 things. Tell how many there are in two parts, 10 ones and some more ones.

> It's easy to find the 10 ones when they are in 5-groups.

> I circle 10 ones and count the rest. Here are 4 more ones.

I have 10 ones and 4 ones.

> It's a little tricky to find the 10 ones here. I make a line so that I remember where I start counting and then keep going around until I get to 10.

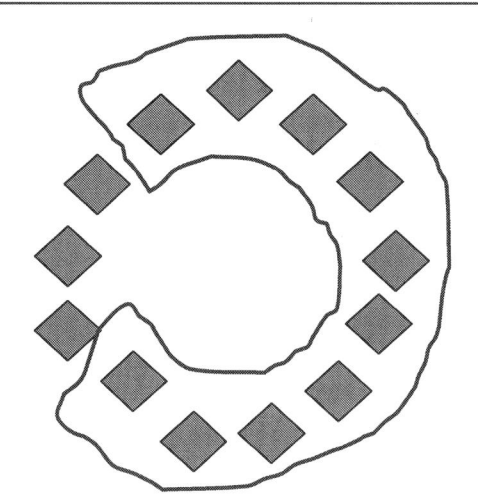

I have 10 ones and 3 ones.

Lesson 3: Count and circle 10 objects within images of 10 to 20 objects, and describe as 10 ones and __ ones.

GK-M5-Lesson 4

Draw a line to match each picture with the numbers the Say Ten way.

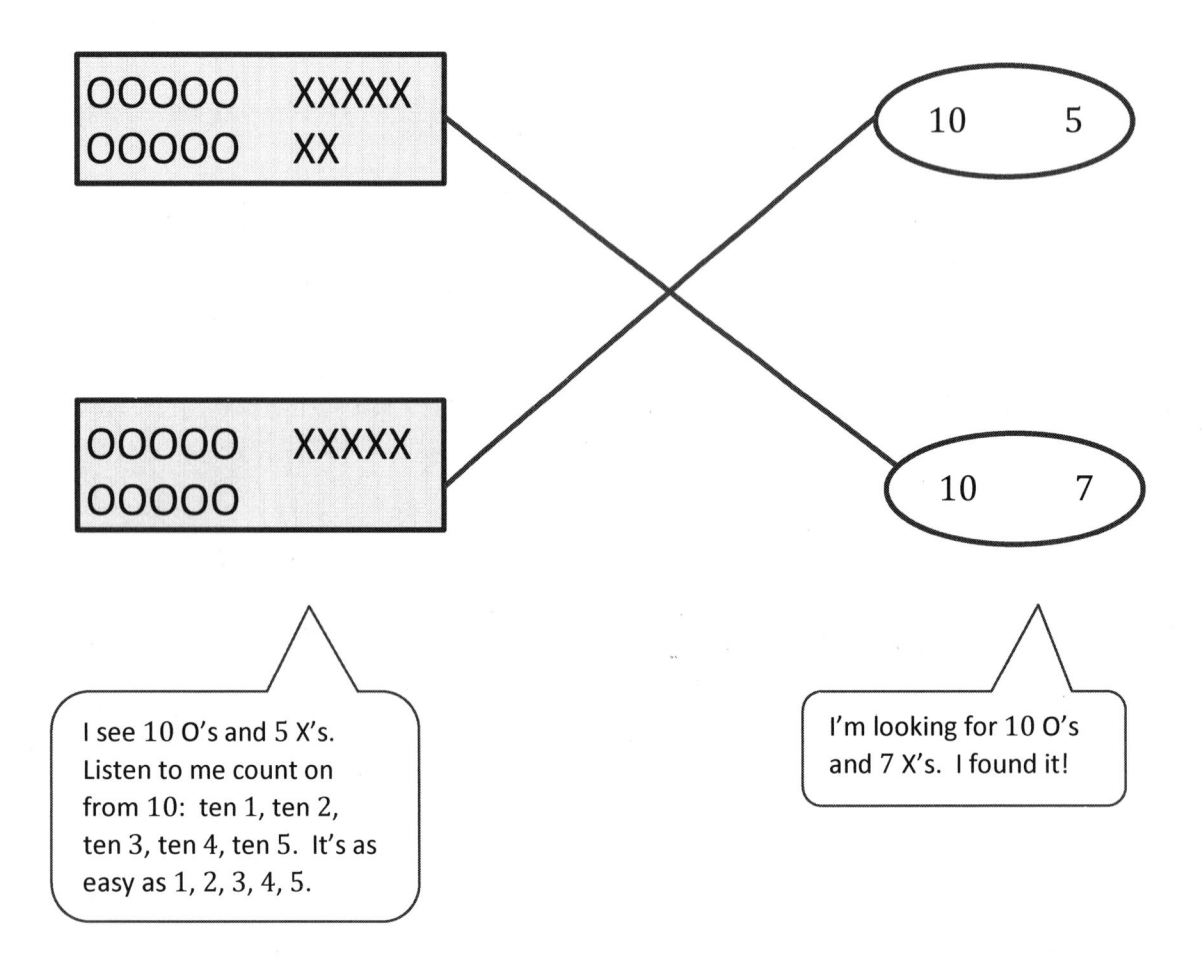

OOOOO XXXXX
OOOOO XX

OOOOO XXXXX
OOOOO

10   5

10   7

I see 10 O's and 5 X's. Listen to me count on from 10: ten 1, ten 2, ten 3, ten 4, ten 5. It's as easy as 1, 2, 3, 4, 5.

I'm looking for 10 O's and 7 X's. I found it!

GK-M5-Lesson 5

Write the numbers that go before and after, counting the Say Ten way.

> Putting "and" in the middle helps me think of the number in two parts.

> I can count the Say Ten way: ten 1, ten 2, ten 3, ten 4, ten 5, ten 6, ten 7, ten 8, ten 9, 2 ten. Another way to say 2 ten is 10 and 10.

BEFORE	NUMBER	AFTER
10 and 2	10 and 3	10 and 4
10 and 6	10 and 7	10 and 8
10 and 7	10 and 8	10 and 9

> I just count the Say Ten way and listen for the numbers before and after. Then I know what to write!

Lesson 5: Count straws the Say Ten way to 20; make a pile for each ten.

GK-M5-Lesson 6

Write and draw the number. Use your Hide Zero cards to help you.

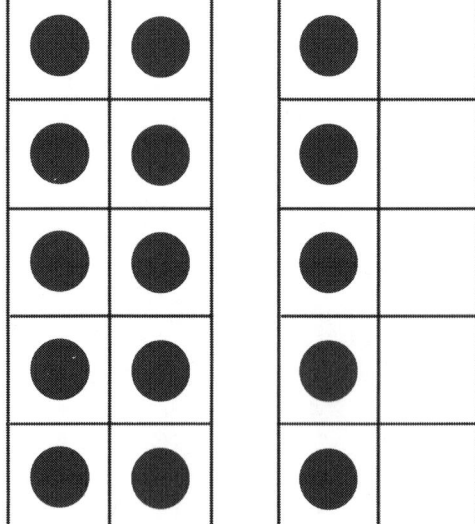

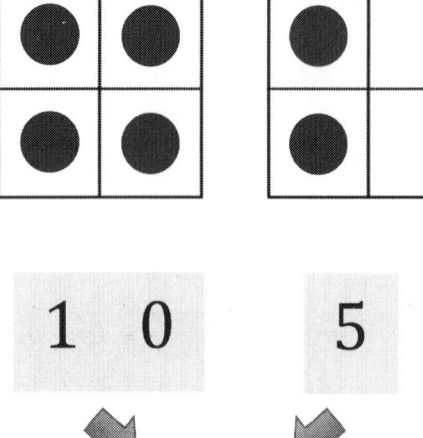

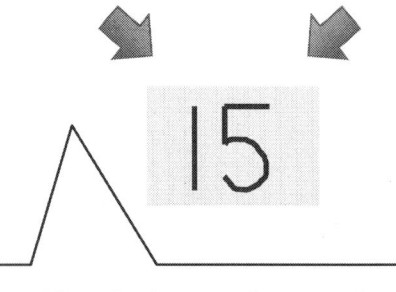

1 0   5

↘ ↙

15

> I can fill in the first ten-frame with dots to show 10 and draw 5 in the next ten-frame. I use my Hide Zero cards to cover the zero in 10 with 5 and see that 10 and 5 makes 15.

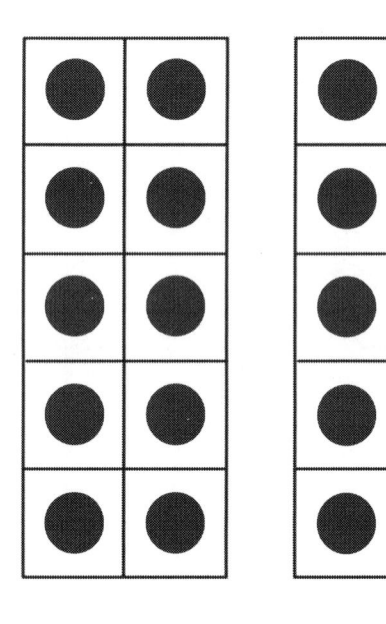

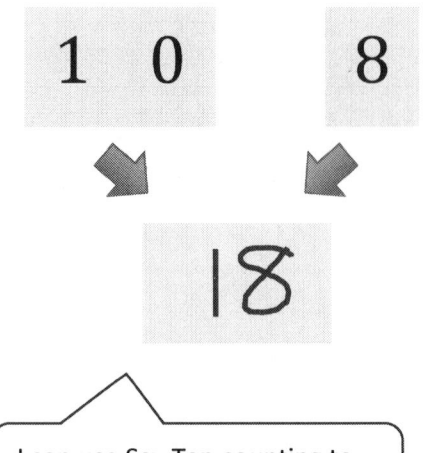

1 0   8

↘ ↙

18

> I can use Say Ten counting to help me. I know ten 8 is 18.

GK-M5-Lesson 7

Look at the Hide Zero cards or the 5-group cards. Use your cards to show the number. Write the number as a number bond.

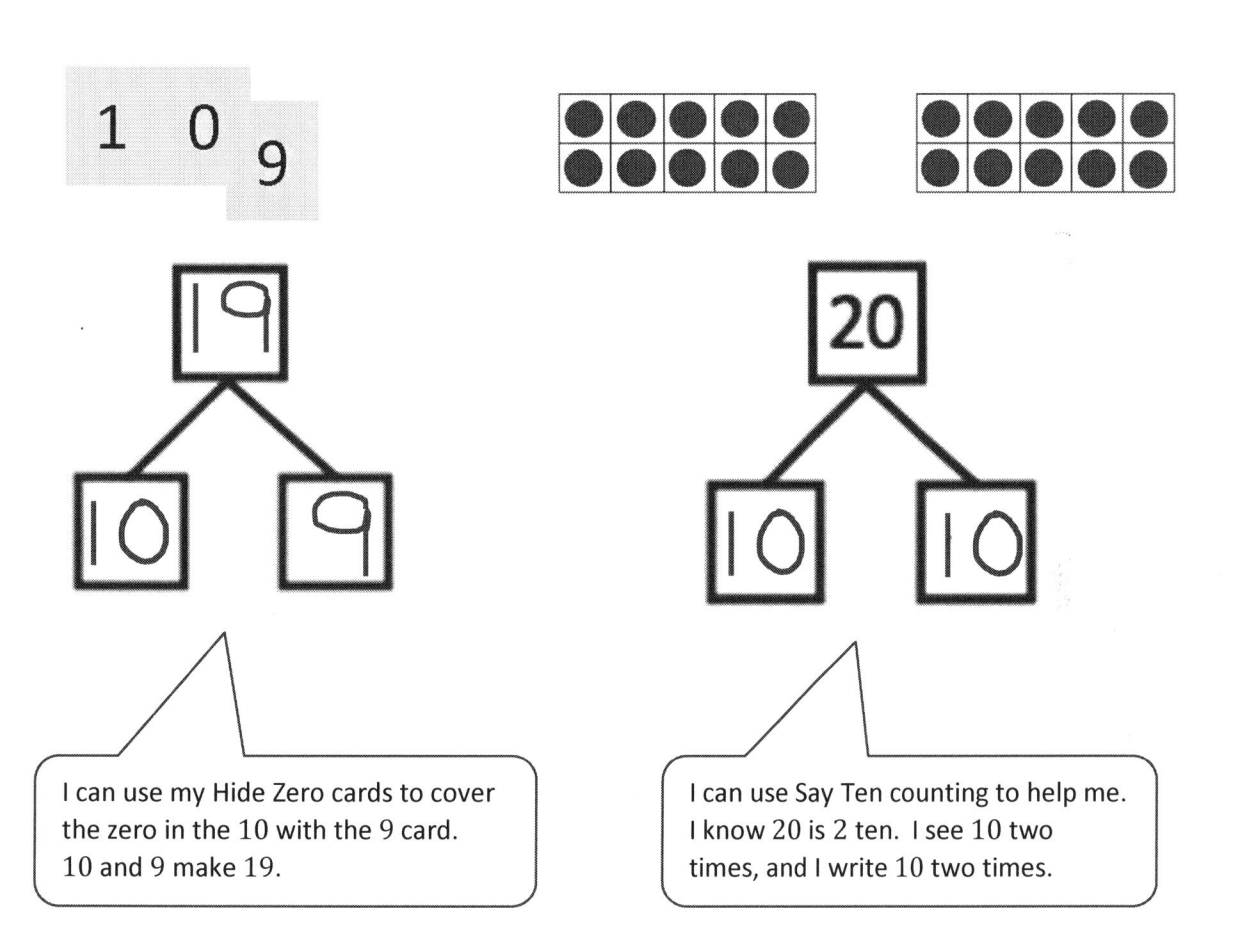

> I can use my Hide Zero cards to cover the zero in the 10 with the 9 card. 10 and 9 make 19.

> I can use Say Ten counting to help me. I know 20 is 2 ten. I see 10 two times, and I write 10 two times.

Lesson 7: Model and write numbers 10 to 20 as number bonds.

Homework Helper

A Story of Units

GK-M5-Lesson 8

Use your materials to show each number as 10 ones and some more ones. Use your 5-groups way of drawing.

| 1   4 |

| Ten six |

> I know 14 is 10 and 4. I can use pennies to show 14. I put down 10 pennies the 5-group way. That's easy. 5 and 5 makes 10. Then I just put 4 more. I can draw a picture of my pennies.

> Ten six is the Say Ten counting way to say 16. This time I can use cereal to show 16. I can draw 16 circles to show how I arrange my o-shaped cereal. I see 10 ones and 6 more ones. I count them like this: ten 1, ten 2, ten 3, ten 4, ten 5, ten 6. I did it right!

Lesson 8: Model teen numbers with materials from abstract to concrete.

Homework Helper   A Story of Units

GK-M5-Lesson 9

For each number, make a drawing that shows that many objects. Circle 10 ones.

13

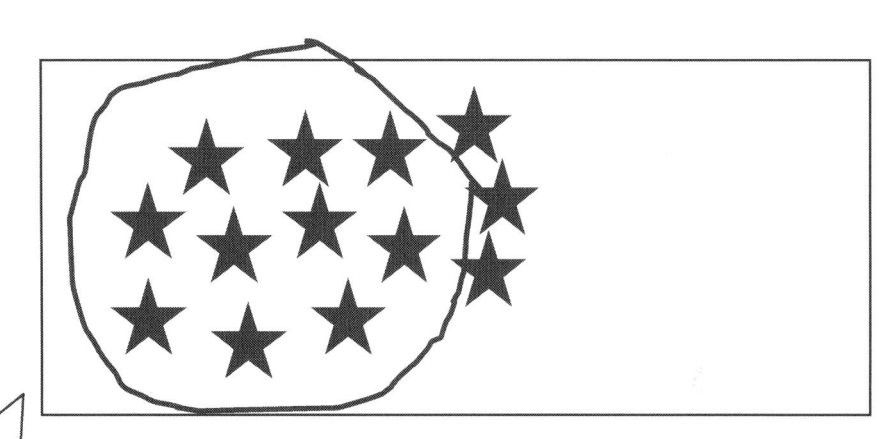

> I can draw 13 stars. I can think of my Hide Zero cards to help me. 13 is 10 ones and 3 more ones.

17

> 17 the Say Ten way is ten 7. I can draw 17 moons in 5-groups to help me see 10 ones and 7 more ones easily.

Lesson 9:   Draw teen numbers from abstract to pictorial.

Homework Helper

A Story of Units

GK-M5-Lesson 10

Color the number of fingernails and beads to match the number bond. Show by coloring 10 ones above and extra ones below. Fill in the number bonds.

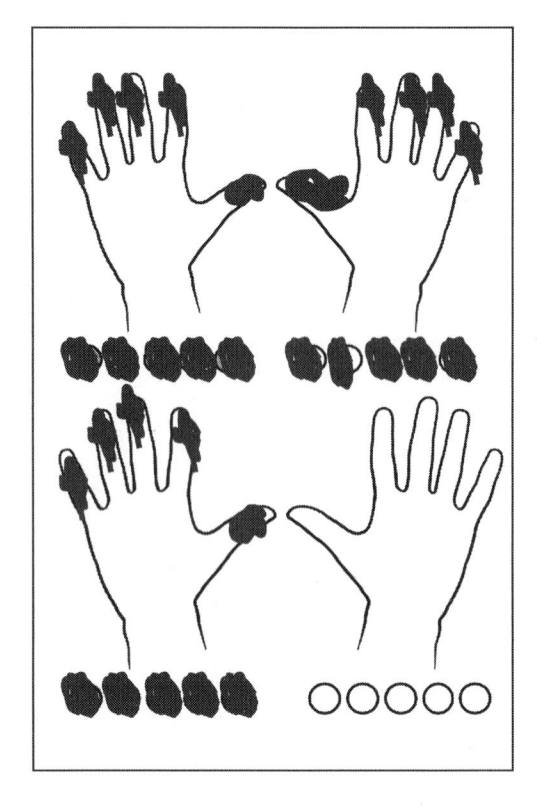

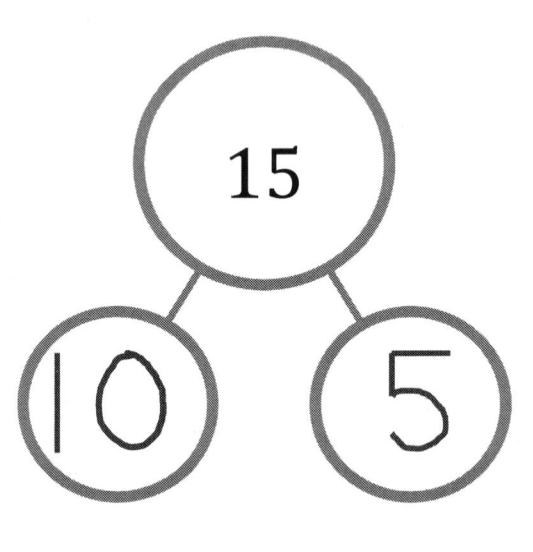

> I know 15 is 10 ones and 5 ones. I can color 10 fingernails and beads on top. I can color 5 more fingernails and beads below. I fill in the number bond with 10 and 5 to match my drawing.

Lesson 10: Build a Rekenrek to 20.

Homework Helper — A Story of Units

GK-M5-Lesson 11

Write the missing numbers. Then, count and draw X's and O's to complete the pattern.

| 10 | 11 | 12 | 13 | 14 | 15 | 16 | 17 | 18 | 19 | 20 |

> To find the missing number, I use the pattern of 1 larger. It goes like this:
> 10. 1 more is 11.
> 11. 1 more is 12.
> I draw 10 O's and 2 X's. Ten 2 is the same as 12.

> I can think of my Hide Zero cards and Say Ten counting, too. I know 19 is 10 ones and 9 more ones. I can draw 10 O's and 9 X's.

EUREKA MATH

Lesson 11: Show, count, and write numbers 11 to 20 in tower configurations increasing by 1—a pattern of *1 larger*.

11

Homework Helper   A Story of Units   K•5

GK-M5-Lesson 12

Write the missing numbers. Then, draw X's and O's to complete the pattern.

20	19	18	17	16	15	14	13	12	11	10

> I count the O's and X's. There are 10 O's and 10 X's. That's 2 ten. 2 ten is the same as 20.

> I know I'm on the right track because I hear the pattern of 1 smaller. It goes like this:
> 14. 1 less is 13.
> 13. 1 less is 12.
> 12. 1 less is 11.

Lesson 12: Represent numbers 20 to 11 in tower configurations decreasing by 1—a pattern of 1 smaller.

GK-M5-Lesson 13

Count the objects. Draw dots to show the same number on the double 10-frames.

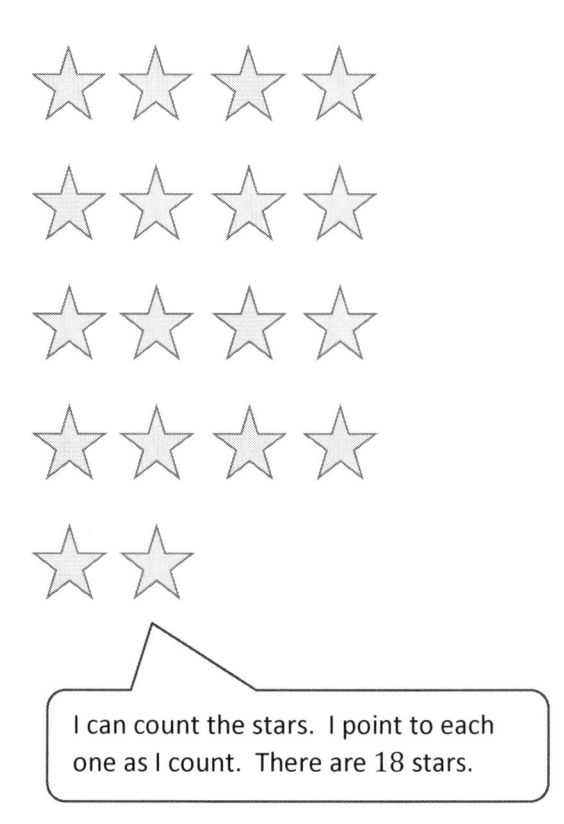

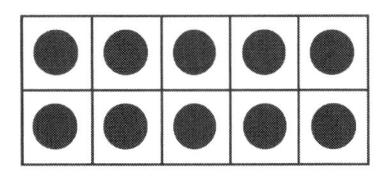

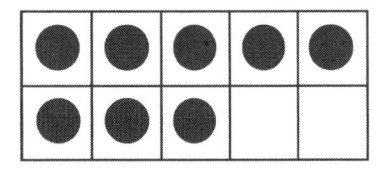

> I can count the stars. I point to each one as I count. There are 18 stars.

> I know 18 the Say Ten way is ten 8. I can fill in the top frame with ten ones and draw 8 more ones in the bottom ten-frame. I can draw 8 ones easily. I know 8 is five and three.

Lesson 13: Show, count, and write to answer *how many* questions in linear and array configurations.

Homework Helper

GK-M5-Lesson 14

Count the objects. Write the number in the box next to the picture.

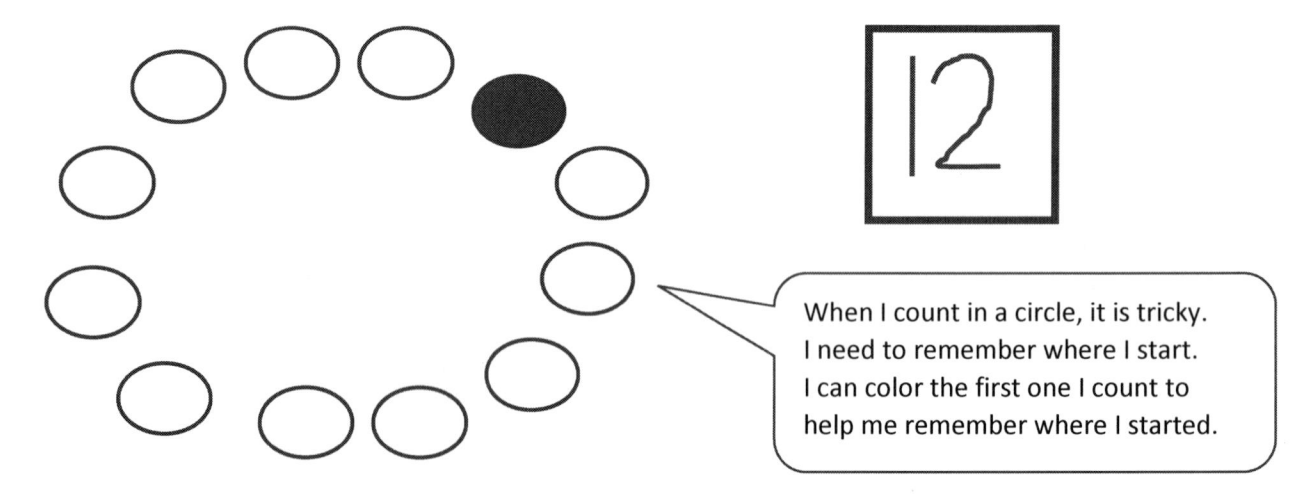

12

> When I count in a circle, it is tricky. I need to remember where I start. I can color the first one I count to help me remember where I started.

Count and draw in more shapes to match the number.

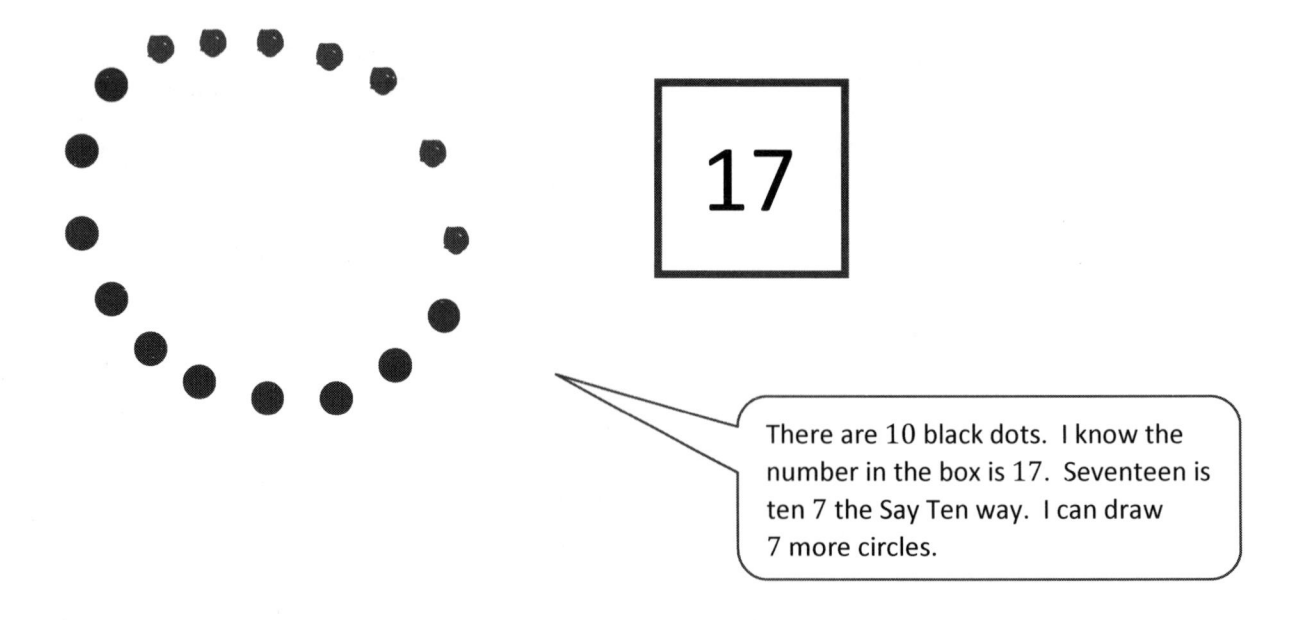

17

> There are 10 black dots. I know the number in the box is 17. Seventeen is ten 7 the Say Ten way. I can draw 7 more circles.

Lesson 14: Show, count, and write to answer *how many* questions with up to 20 objects in circular configurations.

Count the dots. Draw each dot in the 10-frame. Write the number in the box below the 10-frames.

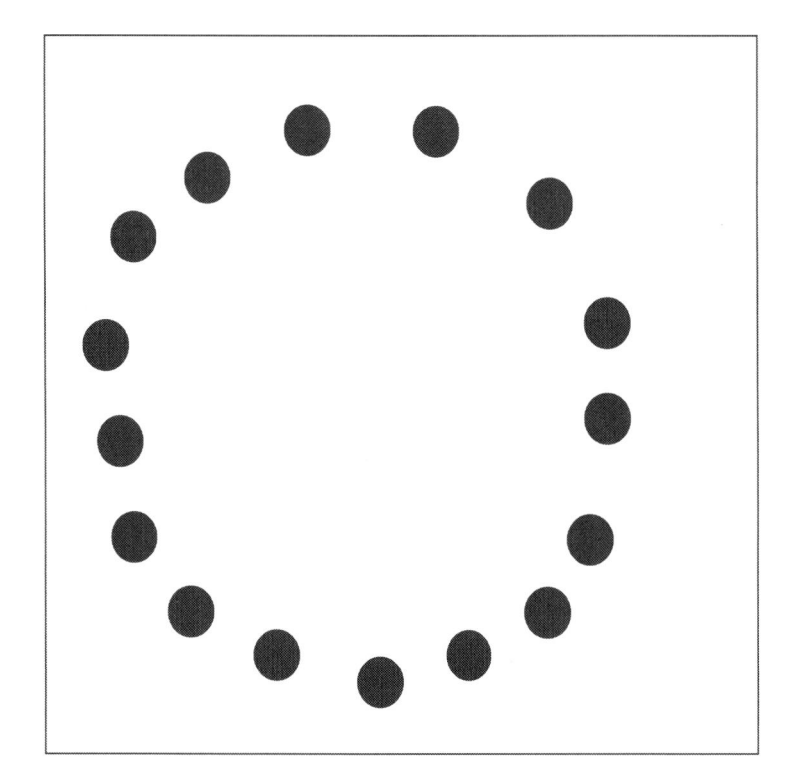

> There are 16 dots. I can draw 16 in the double 10-frames. I can draw 10 in the top frame and draw 6 more in the bottom frame. Sixteen the Say Ten way is ten 6.

16

Lesson 14: Show, count, and write to answer *how many* questions with up to 20 objects in circular configurations.

GK-M5-Lesson 15

Count the Say Ten way. Write the missing numbers.

[6 ten-frames]	60	_6_ tens
[7 ten-frames]	70	7 tens
[8 ten-frames]	80	_8_ tens
[7 ten-frames]	70	_7_ tens
[6 ten-frames]	60	6 tens

> I can count by tens and the Say Ten way! I count the ten-frames first. There are 6 ten-frames, so that is 6 tens. 6 tens is the same as 60.

Homework Helper

GK-M5-Lesson 16

Help the rabbit get his carrot. Count by 1's.

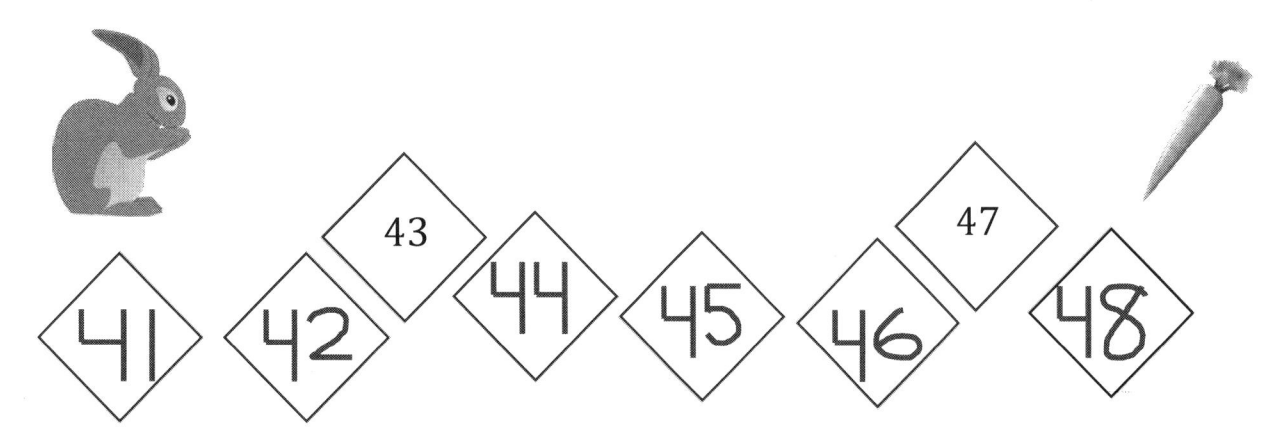

41 > 42 > 43 > 44 > 45 > 46 > 47 > 48

> I help the rabbit get to the carrot by counting by 1's. I count backward from 43 to fill in 42 and 41. Then, I count forward from 43 to fill in the rest of the numbers.

Count up by 1's and then down by 1's.

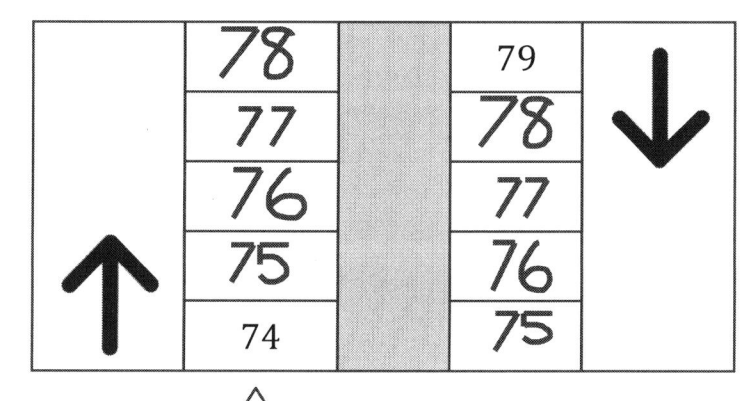

↑	78		79	↓
	77		78	
	76		77	
	75		76	
	74		75	

> I count up starting with 74. Then, I count down in the next column from 79.

Lesson 16: Count within tens by ones.

Homework Helper A Story of Units K•5

GK-M5-Lesson 17

Draw more to show the number.

> 42 is the same as 4 tens 2. The first ten-frame is full, so I don't need to draw more dots. I make dots in each ten-frame until 4 ten-frames are full. Then, I add two more dots to make 42.

42

21

> I draw more dots to get to 20 and then add 1 more to make 21 dots!

Lesson 17: Count across tens when counting by ones through 40.

GK-M5-Lesson 18

Use your Rekenrek, hiding paper (a blank sheet of paper), and crayons to complete each step listed below. Read and complete the problems with the help of an adult.

Hide to show just 40 on your Rekenrek dot paper. Touch and count the circles until you say 24. Color 24 (the 24th circle) green.

- Touch and count each circle from 24 to 32.
- Color 32 (the 32nd circle) with a red crayon.

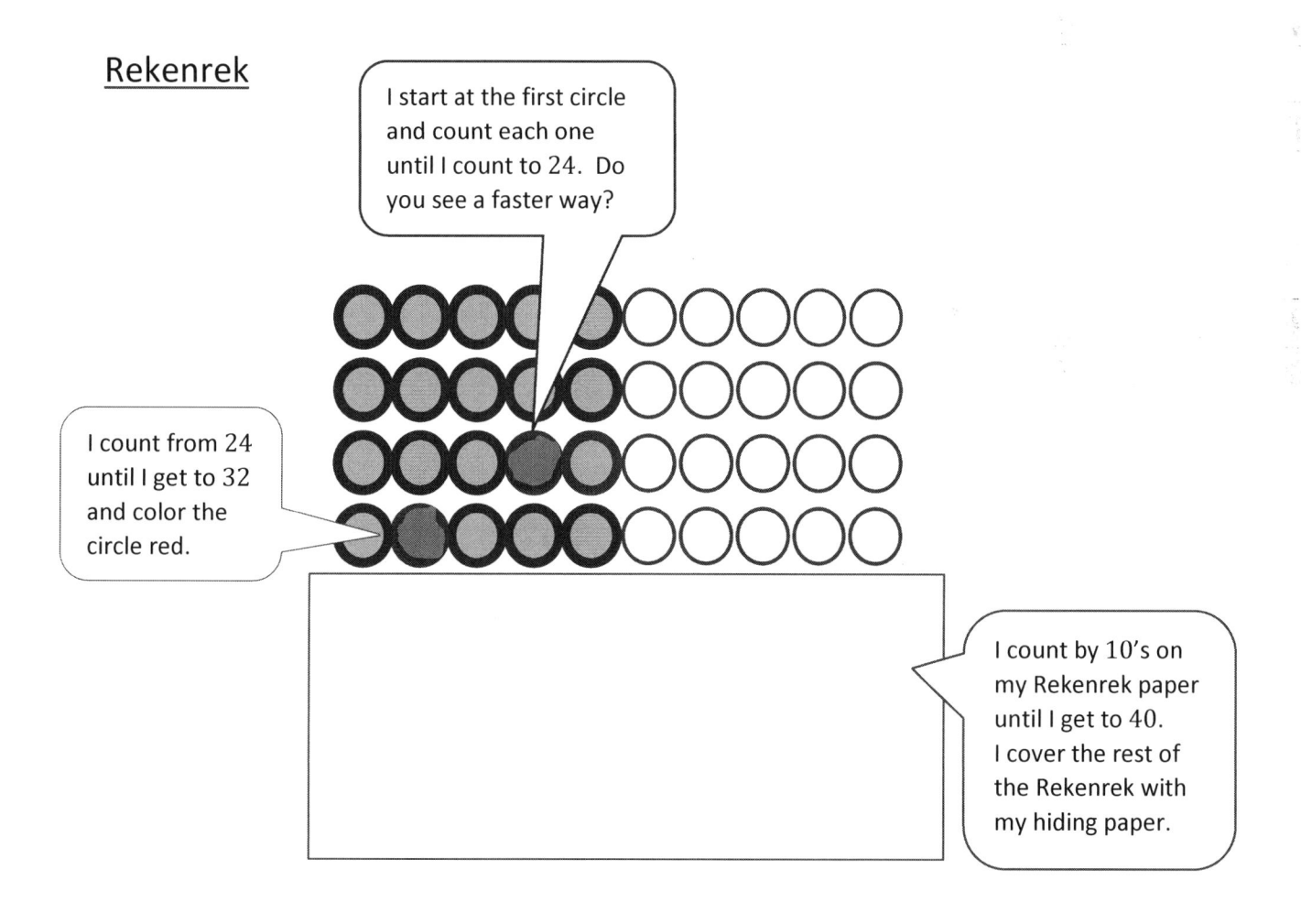

Rekenrek

I start at the first circle and count each one until I count to 24. Do you see a faster way?

I count from 24 until I get to 32 and color the circle red.

I count by 10's on my Rekenrek paper until I get to 40. I cover the rest of the Rekenrek with my hiding paper.

Lesson 18: Count across tens by ones to 100 with and without objects.

Homework Helper

GK-M5-Lesson 19

Write the number you see. Now, draw one more. Then write the new number.

30   31

> I count 30 smiley faces. I draw 1 more smiley face, and now there are 31 smiley faces.

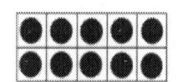

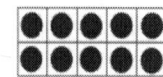

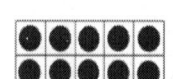

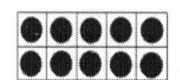

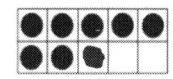

47   48

> I see 4 full ten-frames and 7 dots. That is 4 tens 7. I add a dot, and now there are 4 tens 8, which is 48.

GK-M5-Lesson 20

Draw stars to show the number as a number bond of 10 ones and some ones. Show each example as two addition sentences of 10 ones and some ones.

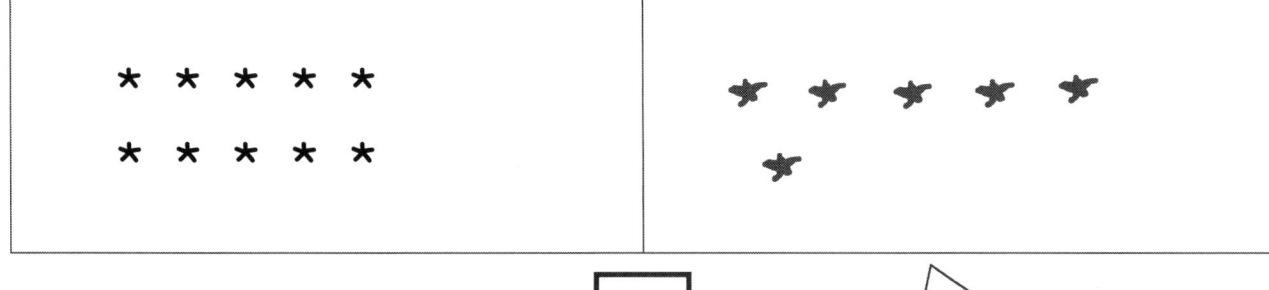

```
* * * * *        * * * * *
* * * * *          *
```

16

> I need to show 16 stars! There are 10 stars, so I draw 6 more to show my two parts.

$$10 + 6 = 16$$

$$16 = 10 + 6$$

> I can make two addition sentences! I show my two parts in the first addition sentence. For the second number sentence, I show the whole first and then the parts.

Lesson 20: Represent teen number compositions and decompositions as addition sentences.

Homework Helper — A Story of Units

GK-M5-Lesson 21

Complete the number bond and number sentence. Draw the cubes of the missing part.

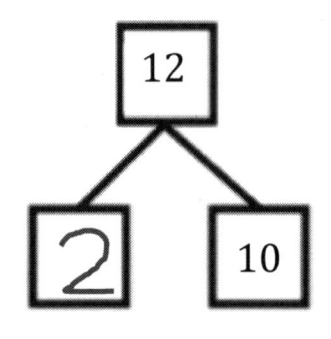

Number bond: 12 = 2 and 10

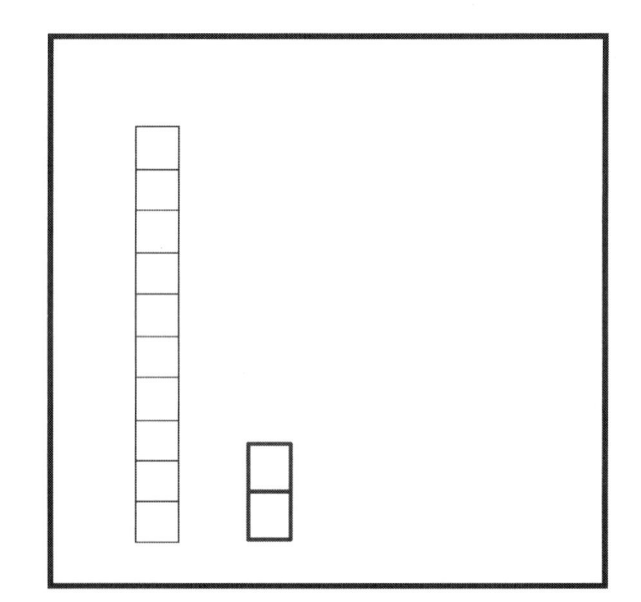

$12 = \underline{2} + 10$

> Hiding inside of 12 are 10 ones and 2 ones. I write a 2 to finish the number bond and the number sentence. There are 10 cubes already there, so I draw 2 more cubes to make ten 2, or 12 cubes.

Lesson 21: Represent teen number decompositions as 10 ones and some ones, and find a hidden part

Homework Helper — A Story of Units

GK-M5-Lesson 22

Fill in the number bond. Check the group with more.

> 14 the Say Ten way is ten 4. I write 4 to finish the missing part of the number bond.

14 → 10 and 4

XXXXX
XXXXX
OOOO

> 18 the Say Ten way is ten 8. I write 10 to finish the missing part of the number bond.

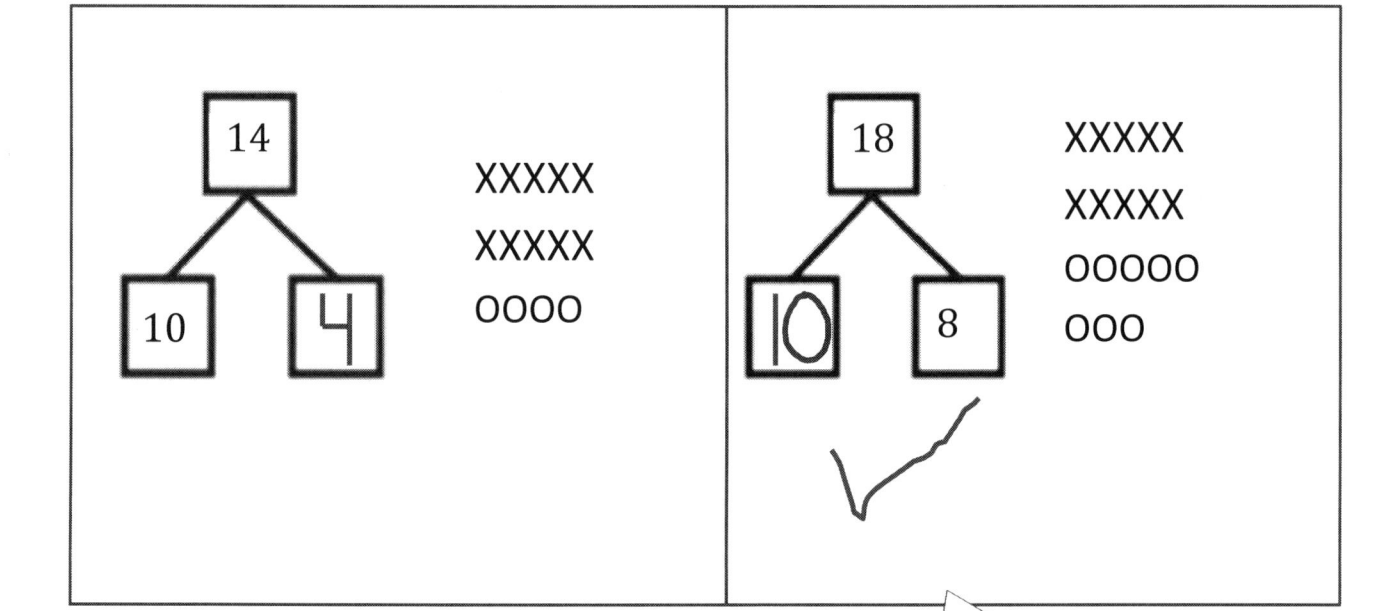

18 → 10 and 8

XXXXX
XXXXX
OOOOO
OOO

✓

> I know both numbers have 10 ones. So, I look at the extra ones to see which has more. 8 is more than 4, so that means ten 8 is more than ten 4.

Lesson 22: Decompose teen numbers as 10 ones and some ones; compare some ones to compare the teen numbers.

GK-M5-Lesson 23

Bob bought 5 strawberry doughnuts and 10 chocolate doughnuts. Draw and show all of Bob's doughnuts.

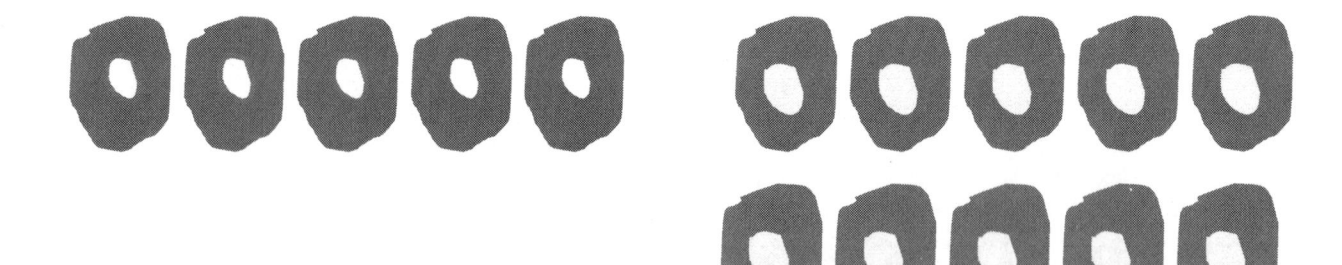

Write an addition sentence to match your drawing.

$$5 + 10 = 15$$

> It's easy to see the doughnuts in two parts: strawberry and chocolate! 5 and 10 is the same as ten 5. That's 15.

> I am great at making addition sentences! Let me tell you how my addition sentence matches my picture. The number 5 tells about the strawberry doughnuts. The number 10 tells about the chocolate. The number 15 tells how many doughnuts in all.

Homework Helper — A Story of Units K•5

GK-M5-Lesson 24

Rabbit and Froggy's Matching Race

Directions: Play Rabbit and Froggy's Matching Race with a friend, relative, or parent to help your animal reach its food first! The first animal to reach the food wins.

- Put your teen numeral and dot cards face down in rows with teen numbers in one row and dot cards in another row.
- Flip to find 2 cards that match.
  Place cards back in the same place if they don't match.
  Continue until you find a match.

- Write a number bond to match. Hop 1 space if you get it right!
- Write a number sentence. Hop 1 space if you get it right!

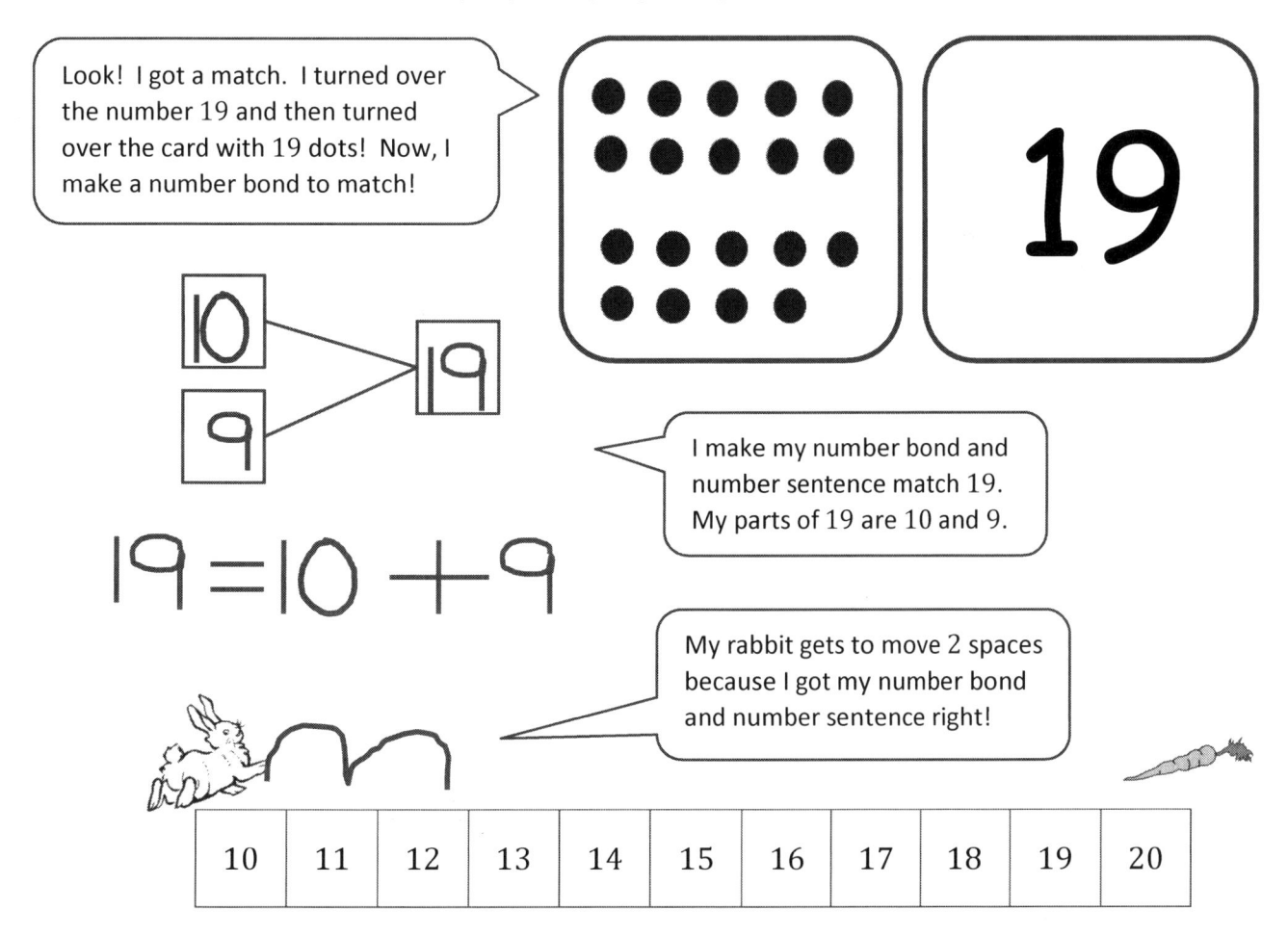

> Look! I got a match. I turned over the number 19 and then turned over the card with 19 dots! Now, I make a number bond to match!

$19 = 10 + 9$

> I make my number bond and number sentence match 19. My parts of 19 are 10 and 9.

> My rabbit gets to move 2 spaces because I got my number bond and number sentence right!

| 10 | 11 | 12 | 13 | 14 | 15 | 16 | 17 | 18 | 19 | 20 |

Lesson 24: Culminating Task—Represent teen number decompositions in various ways.

Homework Helpers

Grade K
Module 6

GK-M6-Lesson 1

First, use your ruler to draw 2 lines to make a square. Second, color the corners red. Third, draw another square.

> I can follow directions! I use my ruler to draw 2 lines to finish the square. Then, I color the corners red.

> I can make a square! A square has 4 straight sides. I work hard to make sure the sides are all the same length.

Lesson 1: Describe the systematic construction of flat shapes using ordinal numbers.

GK-M6-Lesson 2

Trace the shapes. Then, use a ruler to draw similar shapes in the large rectangle.

It is easy to trace shapes! I take my time and try to stay on the dashed line!

Hexagons are tricky to draw because they have 6 sides. The sides don't have to be the same length. I know that as long as the shape is closed and has 6 sides, it is a hexagon!

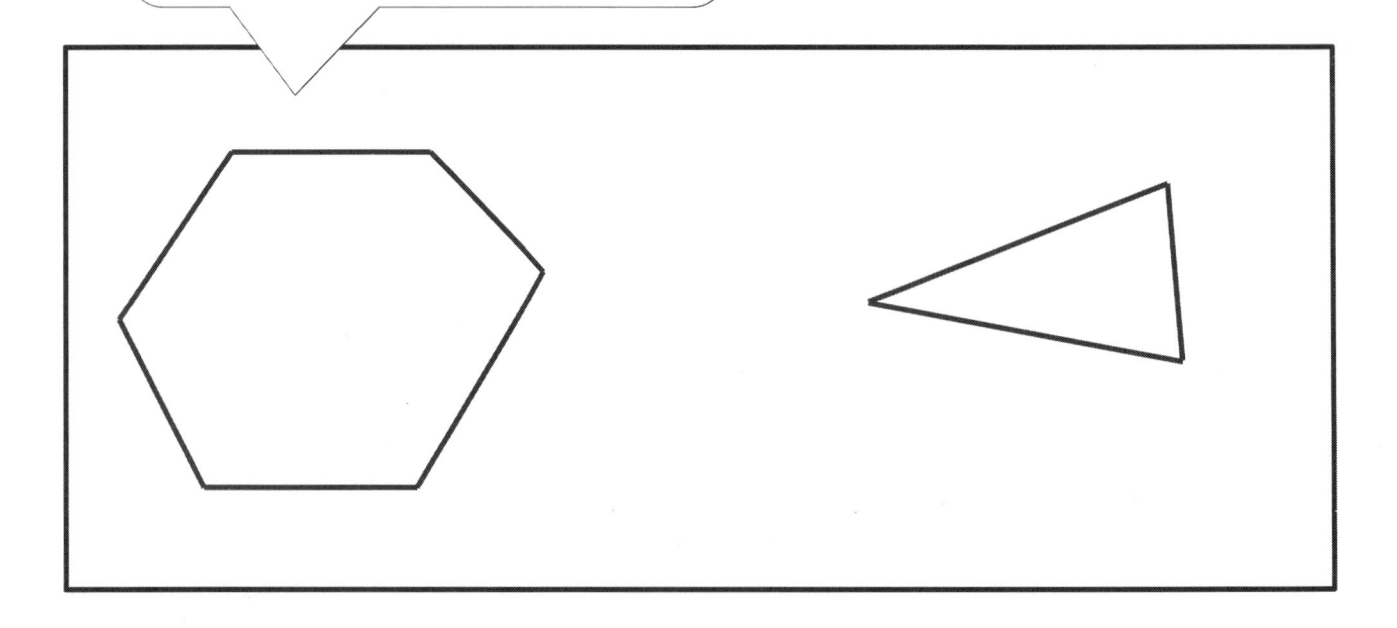

Lesson 2: Build flat shapes with varying side lengths and record with drawings.

Homework Helper — A Story of Units

GK-M6-Lesson 3

Draw something that is a cube.

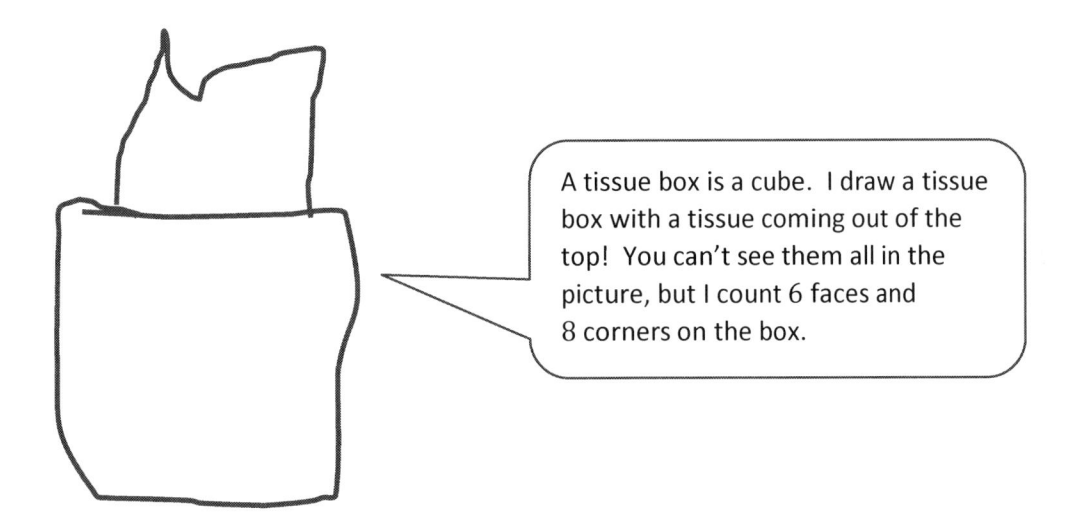

> A tissue box is a cube. I draw a tissue box with a tissue coming out of the top! You can't see them all in the picture, but I count 6 faces and 8 corners on the box.

Circle the flat shape you can see in a ☐ .

> I see squares on a cube! A cube has 6 square faces.

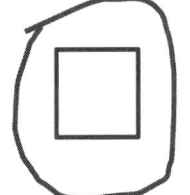

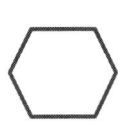

Lesson 3: Compose solids using flat shapes as a foundation.

GK-M6-Lesson 4

Color the 2ⁿᵈ ☆ red.

Color the 4ᵗʰ ☆ blue.

Color the 6ᵗʰ ☆ green.

> The star next to the arrow is the 1ˢᵗ star. That's where I start counting.

→ ☆ ★ ☆ ★ ☆ ★ ☆ ☆ ☆ ☆

> I color the 2ⁿᵈ star red. It is easy to find the second star! I just count 2 stars. I do the same thing with the 4ᵗʰ star.

> I can count to 6 to find the 6ᵗʰ star. Or, I can just count on from the blue one, like this: fooouuur, 5, 6.

Lesson 4: Describe the relative position of shapes using ordinal numbers.

Homework Helper   A Story of Units   K•6

GK-M6-Lesson 5

Match each group of shapes on the left with the new shape they make when they are put together.

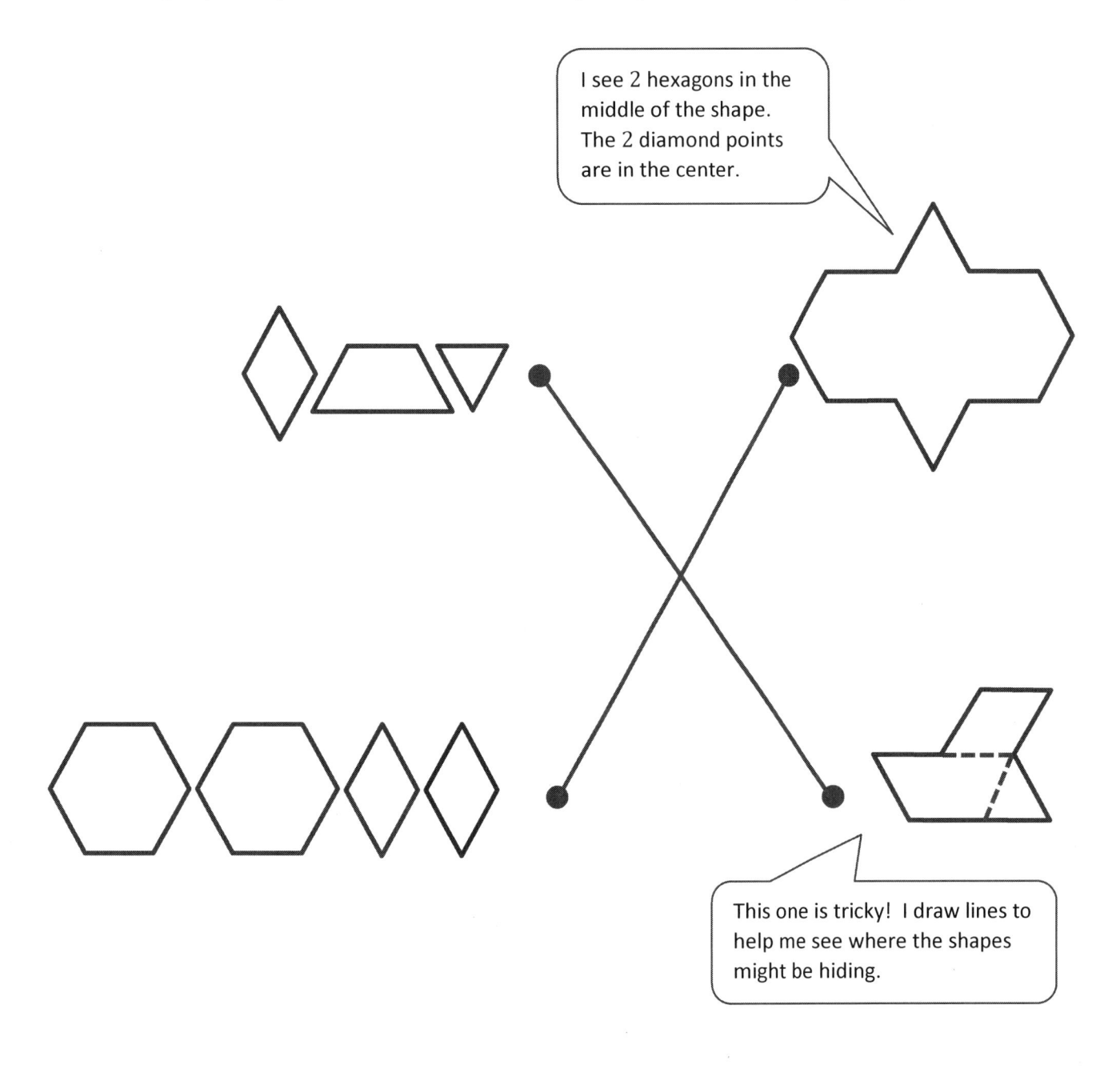

> I see 2 hexagons in the middle of the shape. The 2 diamond points are in the center.

> This one is tricky! I draw lines to help me see where the shapes might be hiding.

Lesson 5: Compose flat shapes using pattern blocks and drawings.

Homework Helper — A Story of Units K•6

GK-M6-Lesson 6

Cut out the triangles at the bottom of the paper. Use the small triangles to make the big shape. Draw lines to show where the triangles fit. Count how many small triangles you used to make the big shape.

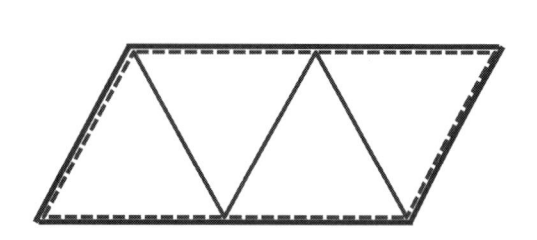

> I use 4 of the triangles to make the big shape. I turn them different ways to make them fit. Then, I trace them. It's like the 4 triangles are hiding inside of the big shape!

The big shape is made with __4__ small triangles.

Lesson 6: Decompose flat shapes into two or more shapes.

GK-M6-Lesson 7

Using your ruler, draw 2 straight lines from side to side through the shape. Describe to an adult the new shapes you made.

First, I make a straight line across the square. Then, I make another line going from the top of the square to the bottom.

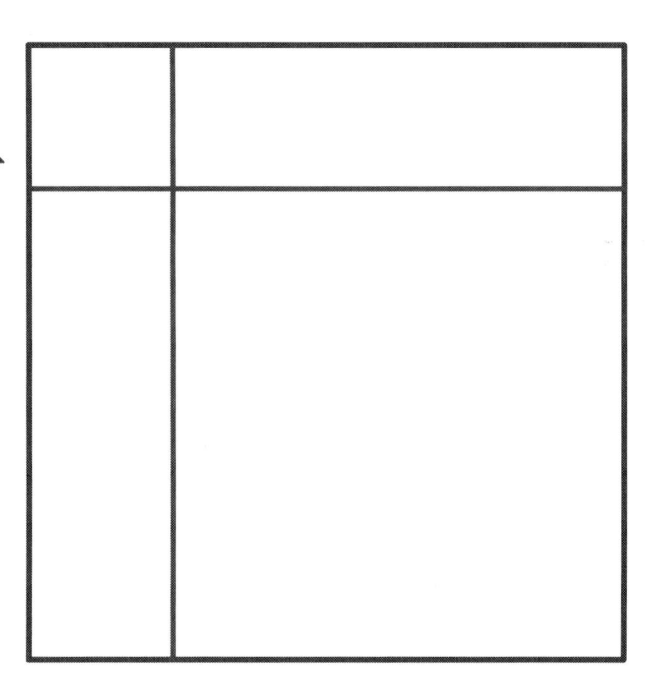

The lines I draw on the square make 4 new rectangles. 2 of the rectangles are squares! It is fun making new shapes!